Of Time and Place

In honour of Oskar Hermann Khristian Spate

O.H.K. Spate, M.A., Ph.D. (Cantab.), FAHA, FASSA

Of Time and Place

Essays in honour of OHK Spate

Editors: JN Jennings and GJR Linge

Australian National University Press 1980
Canberra Australia London England Trumbull Conn. USA

First published in Australia 1980

Printed in Australia for the Australian National University Press, Canberra.

National Library of Australia
Cataloguing-in-Publication entry

Of Time and Place

Index
Bibliography
ISBN 0 7081 1453 9

1. Geography — Addresses, essays, lectures. I. Jennings, Joseph Newell, joint ed. II. Linge, Godfrey James Rutherford, joint ed.

910

Library of Congress No. 79-52350

United Kingdom, Europe, Middle East, and Africa: Books Australia, 3 Henrietta St, London WC2E 8LU, England.
North America: Books Australia, Norwalk, Conn., USA.
Southeast Asia: Angus & Robertson (S.E. Asia) Pty Ltd, Singapore.
Japan: United Publishers Services Ltd, Tokyo.

Phototypesetting and photocomposition by the Federal Capital Press of Australia Pty Ltd. 18 Mort Street, Braddon, 2601, Canberra, Australia.

Printed at Griffin Press Limited, Netley, S.A.

Editors' Preface

In 1951 Oskar Spate committed himself wholeheartedly to Australia and its neighbours; since then his allegiance to The Australian National University has likewise been absolute. It seemed appropriate that invitations to contribute to a book in his honour should be sent only to past and present members of the old Department of Geography in the Research School of Pacific Studies, of which he was the only permanent Head, and of the two successor Departments, Biogeography and Geomorphology and Human Geography, of which he was truly the progenitor.

Economics dictated that this volume should be modest in size and there had to be a thinning to less than half the offers of contributions. We express our regrets to those friends and colleagues of Spate all over the world who would have wished to join in this token of respect but in the one way or the other were not accorded that opportunity.

To have been able to accept more writings would have made it easier to have produced an offering which was representative of the great diversity of interests of the man to whom it is dedicated. Perforce a restricted number of themes was suggested to the people invited to indicate their wish to participate. In the upshot not all these themes survived but it was subject matter alone which guided our choice from the offers made. In this there was an element of egalitarianism not inapt given Oskar Spate's *Weltanschauung*. The regional basis of the material incorporated — in effect the realm of interest of the Research School of Pacific Studies of which he is a former Director and still a member — perhaps sets only bounds to the book, rather than provides a cohering theme. There is, however, a chain of connection from one essay to the next with nodes of interest. Even that claim may be thought to place it with advantage amongst the lists of *Festschriften* so many of which seem to have been directed by enthusiasm and sense of duty rather than by critical thought about structure.

To have been able to accept all the contributions offered might have rendered the choice of title for the assemblage even more difficult than it has proven. We asked for suggestions; to mention some of them here is not without point. 'A Spate of Oscars' was no doubt prompted by familiarity with his lifelong motto that 'one doesn't have to be solemn to be serious'. But with the wit of this proposal, characteristic of Spate, goes a pretension which he would be the first to prick.

On similar grounds we had to reject a suggestion otherwise excellent, 'The Science of Princes', a title which would have at least brought it on to many a geographer's coffee-table, if not to his study. This Johnsonian quotation may escape some of our readers as it did ourselves so we place it in context here:

> Geography is in a peculiar manner the science of Princes. When a private student revolves the terraqueous globe, he beholds a succession of countries in which he has no more interest than in the imaginary regions of Jupiter and Saturn. But Your Majesty must contemplate the scientific picture with other sentiments, and consider, as oceans and continents are rolling before You, how large a part of mankind is now waiting on Your determinations, and may receive benefits or suffer evils, as Your influence is extended or withdrawn.

Another reason for shunning this title is that it inevitably calls up political geography, a prime interest of Spate's which self exile in the calm Antipodes has unfortunately but inevitably stilled to some degree.

'A Silken Purse' conveys nicely the notion of a gift both of worth and elegance and its royal or noble flavour does not now seem overly presumptuous. Yet today it is best known through a popular saying with unfortunate porcine associations, too ready a handle to give unkindly reviewers. John Donne, a poet highly esteemed by Spate, yielded 'Spoil of the Old World's Farthest End', appropriate enough for the area studies here encompassed but perhaps a little precious for the book, though not for the most purely academic theme contained within it.

The contents of the book as a whole fit easily into a notion of geography as the study of land-man relationships but without doubt the dominant approach is from the side of man. This is thoroughly in keeping with Spate's own works and this slant suggested as a title a favourite phrase of his which hails from Terence — 'Think Nothing Human Indifferent'. Anyone who knows Spate well cannot escape becoming aware of the fact that nothing matters more to him than the personal fate of the individual human so that there is much to be said for this phrase as a title. Yet a title should be intelligible to a wide audience without too much mind searching, and changing word usage renders this translation somewhat esoteric today.

We regret that we could not decently tap Spate's own knowledge of the minor poets of the seventeenth and eighteenth centuries to which he is addicted and had to turn instead to the words of a Victorian poet — neither his favourite century nor his favourite poet.

> For tho' from out our bourne of Time and Place
> The flood may bear me far . . .

The final choice, 'Of Time and Place', applies equally well to the contents of the book and to Oskar Spate's abiding scholarly interests.

J. N. Jennings
Canberra, 1979 G. J. R. Linge

Editors' Acknowledgments

We thank the contributors to this volume for their co-operation and for their patience in responding to seemingly endless editorial inquiries. Many other people in the Departments of Human Geography and Biogeography and Geomorphology assisted us. We appreciate the help of Grace Richardson who prepared the list of Oskar Spate's writings; of Pauline Falconer, Rowena Barker, Marlene Arney and Jan Zammit who typed and retyped various versions of the manuscript; of Keith Mitchell who prepared the line drawings; and of Ken Lockwood for his assistance with the reproduction of some of the illustrations. We also wish to acknowledge the support and encouragement given by Betty Jennings and Jan Linge.

The Title Page Drawing

The drawing on the title page is a representation of John Harrison's fourth timekeeper completed in 1759. This clock kept sufficiently accurate time on board ship to solve the problem of determining longitude at sea. A replica was made for use aboard *Resolution* during Captain Cook's second voyage (1772-5). On 19 July 1775 it was noted in the log:

We made land about Plymouth, Maker Church, at 5 o'clock in the afternoon. Bore North 100° West distant 4 leagues. This bearing and distance show that the error of the watch in longitude was only 7 minutes 45 seconds, that is, less than 6 miles too far to the West.

Time and place had been resolved.

Contents

Tables

Figures

Oskar Spate

A Personal Impression

*T. M. Perry**

We met on the steps of the Australian Museum in Sydney and sat in Hyde Park, gazing, like the nearby statue of Captain Cook, down the harbour. We had met to discuss the possibility of my becoming a scholar in Spate's Department at The Australian National University; our talk was part interview, part *viva voce* examination and part friendly chat. At its end I felt by no means sure of a scholarship. Nonetheless I was sure that I had found a friend with a lively interest in a vast range of subjects who, in a couple of years, had learned a good deal about Australia and its historical geography, and who would get right to the point. His perceptiveness was not to be evaded or confused. These impressions were confirmed and strengthened by three years in Canberra and during the years that have since passed. He did not say 'all generalisations are dangerous, even this one', but he might have done so: as a mini-Boswell I fear the Johnsonian epithets these words might earn. I trust him to act on his own advice: 'you must accept with outward equanimity the things they will by now be saying about you' (Spate 1965: 7).

Oskar Hermann Kristian Spate was born in England on 30 March 1911, spent some childhood years in the United States of America, and, back in London, attended St Clement Danes

* *T.M. Perry is Reader in Geography at The University of Melbourne. He graduated from Sydney University and took his doctoral degree in the Department of Geography, Research School of Pacific Studies, The Australian National University, in 1958. He was the first Research Scholar of the Department to take the Ph.D. degree.*

Grammar School where Leonard Suggate was one of his teachers. In 1930 he went to Cambridge and St Catharine's College as a Scholar in English and Geography. Despite the disagreement with a Tutor which produced the lines:

> Within this College thought is free
> Provided that you think like me

he won first class honours in Part I of both Triposes, and in Part II of the Geographical Tripos.

At that time there was in Cambridge geography a strong reaction against determinism and environmentalism: Spate, who considered himself a Marxist, was faced with the problem of reconciling dialectical materialism with the possibilist approach to geography advocated by Febvre, Vallaux, Vidal de la Blache and Brunhes. This dilemma provided a theme to which he has repeatedly returned, it is after all (even without political colour) basic to most theorising about geography, and for which he was eventually to suggest a compromise in 'probabilism'. He was impatient of that type of meticulous scholarship which tends to produce in undergraduates 'an absolute terror of drawing any conclusion in case you should also draw a mistake'; but despite his efforts to become an *enfant terrible,* Spatian generalisations and conclusions are firmly founded in fact. Within Spate the theoretician there lurks an empiricist, and his critiques of the work of Wittfogel, Toynbee and Huntington (Spate 1959: 90-5; 1965: 185-233), for example, nail dubious arguments based on broad generalisations with precise geographical knowledge. After obtaining his B.A. he began research on the development of London for his Ph.D. (completed in 1937). He was still a Marxist and after a year at Reading was 'railroaded to Rangoon in order to let [him] mellow down and become a Tory'. He did not. In 1941 he joined the Anti-Aircraft Regiment of the Burma Auxiliary Forces and on Christmas Eve 'after about five and a half minutes of active service' was wounded by one of the first Japanese bombs to fall on Rangoon. A long convalescence was followed by periods as a Military Press Censor in Bombay and as Officer-in-Charge of the Burma Section of the Inter-Services Topographical Department in the South East Asia Command Headquarters. While with the Topographical Department he contemplated writing a regional geography of the Indian subcontinent, an idea that was turned to resolve by an invitation from Methuen to do just that. The writing of this book occupied his post-war years while at Bedford College and then, as Lecturer and Reader, at the London School of Economics. In 1947 he was adviser to the Ahmadiyya Muslim Community on the Punjab Boundary Commission, an experience that gave him an insight on the birth of Pakistan and the personalities involved in it. He was awarded the Gill Memorial prize of the Royal Geographical Society in 1948 for his work on the geography of India and Burma.

He arrived in Australia in 1951 to become Foundation Professor of Geography in the Research School of Pacific Studies in the recently founded Australian National University — then a University devoted entirely to research. He was welcomed with suitably ocker verses beginning

> Here's to you bloody Spates,
> Youse and we are bloody mates . . .

Australian geography, at that time, was just beginning to grow; only The University of Sydney offered a full honours undergraduate course, but there were courses in most of the other universities — foundations on which the rapid development of the 1960s was based. Spate's task was twofold: to establish his own research department and to help foster the development of geography in the other universities (to what extent the developments in the 1960s were the result of his efforts or products of growth already beginning the future must judge; but it cannot be denied that his Department did provide staff for the expanding departments in the State universities).

India and Pakistan was nearing completion. It was published in 1954 and the reviewers were (almost) unanimously laudatory, though they differed on whether it was 'a milestone' or 'a triumphal arch'. (One critic thought that parts were classic and parts read like a knitting pattern; in the second edition the ground rules were simplified and the need to check that one had the differences between N/S and N-S the right way round disappeared.) Whether it is the last of the great one-man regional geographies may be debated. It did break away from the traditions of its predecessors not so much in its broad organisation of the data (the chapter-heads are traditional enough) but in its discussion of the broader, historical, political and social circumstances that are reflected in regional differences within the subcontinent and which have had, and will have, an impact on past development and future plans for development. Thus it embraced a much wider view than was usual in 'general and regional' geographies. *India and Pakistan* brought its author honours: the Charles Garnier Medal of the Société de Géographie of Paris, the Jawaharlal Nehru Medal of The National Geographical Society of India (both in 1956), and a Russian translation published in 1957. The Governments of the subcontinent were not quite so enthusiastic: some of the text figures, though based on officially published maps, were not official enough, and the import of the book was banned at the same time as the Census Commissioner in India was demanding copies for his offices.

Spate's Geography Department at The Australian National University occupied rooms in Canberra's old hospital building, its wards having been divided to provide large rooms for the staff on the western side of a north-south corridor and small rooms on the eastern side for lesser beings. Geography's scholars were fortunate for, unlike those of other departments, they had rooms — the official reason being that they were an untrustworthy lot who could not be allowed to take maps away from the Department and so expose them to damage or loss. When I became a student in 1954 the Department occupied about two-thirds of the eastern wing of the Old Hospital Building (a description had become a name). Joe Jennings manned the southern frontier with History, Oskar and Joan Binns (his long-suffering Secretary) occupied the Heartland (Joan by day, Oskar by night), and Herb King the northern marchlands commanding the route to the tea room. There were already two Scholars, Bob Wilson and Slim Bauer, and soon there were three more and Trevor Langford-Smith arrived to take up a Research Fellowship. At the beginning of first term the Department was still recovering from organising Section P (Geography) of the Australian and New Zealand Association for the Advancement of Science, and the University was recovering from the trauma of the opening of University House (or more correctly the party that followed the opening). The Research Schools of Pacific Studies and Social Sciences were still small enough for everybody to know everybody else (and even some people in the Research Schools of Physical Sciences and Medical Research). Scholars at last outnumbered the academic staff and Australia's newest university was an exciting place to be.

Memories: certainly of hard work but also of lots of fun; of late nights (and early mornings) in the Department when the 'night shift' was invited into Oskar's room for a chat and a nightcap; of the night when I actually worked later than Oskar, much to his chagrin; of rum punch parties, including the one during which a helpful scholar added sugar to the coffee cauldron instead of the punch one (several ladies thought that they had sugared their coffee under the influence of the punch); of my door opening and Oskar saying 'Tom, you don't need supervising this week do you. Good!'; of Oskar coming in, scrutinising the book shelves and then sitting down for long, and helpful talks about my research, and often devising the form-of-words that was eluding me; and of a 'para-geographical' body of which Oskar is Perpetual Past President, that raised its eyebrows at the first excesses of the 'quantitative revolution' in Australia and pioneered work with the evapo-oskarimeter in measuring saturation deficits.

I also remember Oskar's Island that was mapped, and mapped again (it is too erudite to be called a doodle) during seminars. It began with a coast line, then came rivers, contours, ports, towns, railways, roads, forests and fields: a complete topography, a Dibdinish 'Snug little island, A right little, tight little island'. And despite the apparent complete absorption with cartography, he did not miss a word and at the end usually had a comment that went straight to the heart of the matter.

While the Department gradually acquired new Scholars and Research Fellows, Oskar was involved with the revision of *India and Pakistan*, the writing of papers, the inevitable university committees and boards and with broader fields of professional work. In 1953 he was convener of an Australian National University Working Party advising the Minister for Territories on the economic development of New Guinea; in 1956 he organised a seminar on the use of regional surveys in national planning for the Indian Statistical Institute; and in 1958-9 made a survey of the economic problems of the Fijian people for the Government of Fiji. His report, *The Fijian People . . .*, translated into Fijian as *Na Kawa I Taukei . . .*, is, he claims, probably the longest work in that language which is not religious or fiction. His essay 'On Being an Expert' (1965:3-14) obviously owes much to his experiences as a one-man commission of inquiry. He was one of the three Members of the Commission on Higher Education in Papua New Guinea (1963-5) and a member of the 1969 Commission on Education in Fiji; subsequently and consequently, he became a member of the Interim Councils and then of the Councils of The University of Papua and New Guinea and The University of the South Pacific. His work as a geographer was acclaimed by The American Geographical Society's award of the Charles P. Daly Gold Medal in 1969, and by The Royal Geographical Society's Victoria Medal in 1971 for 'outstanding contributions to Indo-Pacific Studies'.

In 1967 he vacated his Chair to become Director of the Research School of Pacific Studies. This was, he has said, partly because of the challenge of a new job and partly that he was beginning to feel out of touch with geography, by which he meant out of touch with the increasing sophistication of research particularly in urban and economic geography in which the complexity of inter-relationships demands the use of statistical techniques to cope with the data, and out of touch with the theorising and model-making that those techniques involved. To the man who ran at the masthead of both his inaugural lecture, and his collection of essays, the lines

> Let me enjoy the earth no less
> Because the all-enacting Might
> That fashioned forth its loveliness
> Had other aims than my delight

from Hardy, who could delight in the architecture and symbolism of Konarak, who could feel the spirit (and spirits?) of places, there is little loveliness or enjoyment in the featureless plain of uniform fertility (even when one river is permitted). When his term as Director of the School ended in 1972 he went to the Department of Pacific History, doubtless to spare his successor in the Chair (now of Human Geography) the embarrassment of an *éminence grise* in the Department. The years before his retirement in 1976 (and since) have been spent writing a history of the Pacific since Magellan — a subject perhaps more complex than the regional geography of India and Pakistan, but replete with thousands of 'right little, tight little' islands, and redolent with the poetry of human vision, heroism and idealism and the prose of human greed, baseness and inhumanity.

The list of publications in this volume betokens the breadth of his interests, and even a cursory examination of almost any one will provide evidence of the depth of his scholarship and reading, not only in geography but also in history and literature (and in languages other than English). His contributions to 'the literature' though largely geographical include items as diverse as significant essays in Australian and Pacific history, and poems translated from the Portuguese. Who but Spate would know of, and cite a quotation from that almost unknown German author of *Uebersicht der Naturwissenschaftlehre Englands in den XVI und XVII Jahrhunderten,* Endorn Weltsz?

If diversity is one of the marks of his writing, style is another: the apt word and the evocative quotation lend force to the argument and command attention. And he does not have to be solemn to be serious, and eschews pretentious jargon.

At the end of his inaugural lecture at The Australian National University Oskar spoke of his aim: 'to make what one can of this our earth, and if the cosmic order holds nothing for us either of despair or hope, to find our happiness in social duty and private love'. In offering this volume to him as a token of our gratitude and respect, we, his students and colleagues, knowing that he has experienced both despair and hope which are but part of life, congratulate him on 'social duty' nobly done and rejoice that private love has brought him happiness and a new enjoyment of the earth — he has at last discovered the joys of a study at home and of gardening.

References

Spate, O.H.K. 1959. The 'Hydraulic Society': K.A. Wittfogel's "Oriental Despotism: a Comparative Study of Total Power" (review article). *Annals of the Association of American Geographers* 49: 90-5.

Spate, O.H.K. 1965. *Let Me Enjoy: Essays, Partly Geographical.* Australian National University Press, Canberra.

Writings of O.H.K. Spate

1934

(with R. Barnes) Barricades on Bankside or, the adventures of two Marxists in searche of a heroe. *Cambridge Left* 1: 81-90.

1936

The growth of London, 1660-1800. Pp. 529-48 in H.C. Darby (ed.), *An Historical Geography of England before A.D. 1800: Fourteen Studies*. Cambridge University Press, Cambridge.

John Strype. *St Catharine's Society Magazine:* 46-50.

1937

Nathaniel Bacon. *St Catharine's Society Magazine:* 73-8.

1938

Geographical aspects of the industrial evolution of London till 1850. *Geographical Journal* 92: 422-32. Reprinted 1976, pp. 275-91 in *Sonderdruck aus Industriegeographie,* Wege der Forschung 329, Wissenschaftliche Buchgesellschaft, Darmstadt.

1941

Beginnings of industrialization in Burma. *Economic Geography* 17: 75-92.

1942

(with L.W. Trueblood) Rangoon: a study in urban geography. *Geographical Review* 32: 56-73.

Factors in the development of capital cities. *Geographical Review* 32: 622-31. Reprinted 1964, pp. 164-71 in W.A.D. Jackson (ed.), *Politics and Geographic Relationships,* Prentice-Hall, New Jersey; and 1967, pp. 413-24 in H.J. de Blij (ed.), *Systematic Political Geography*. Wiley, New York.

1943

Burma Setting. (Burma Pamphlets No. 2.) Longmans, London. Pp. 34.

(with G. Appleton and B.R. Pearn) Burma background. *Current Affairs No. 16*. Pp. 19.

(with G. Appleton and B.R. Pearn) Burma scene. *Current Affairs No. 19*. Pp. 15.

Geographical aspects of the Pakistan scheme. *Geographical Journal* 102: 125-36.

1944

Geography and federalism. *Indian Geographical Journal* 19(1): 24-36.

1945

The Burmese village. *Geographical Review* 35: 523-43.

1947

The resources of Pakistan. *Eastern World* 1(7): 29-30.

The partition of the Punjab and of Bengal. *Geographical Journal* 110: 201-18.

1948

On the regional division of the Indian sub-continent: an introductory essay. *Indian Geographical Journal* 22(2): 1-5.

A year of change: territorial reorganization in the Indian union. *Geographical Journal* 111: 288-92.

The boundary award in the Punjab. *Asiatic Review* January: 1-12.

The partition of India and the prospects of Pakistan. *Geographical Review* 38: 5-29.

The Indo-Pakistan boundaries. *Times Review of Industry* April: 9.

The evolution of the north-west frontier: a study in border policy. *Eastern World* 2(8-9): 12-15.

1949

Integration in India. *Eastern World* 3(7):6-7.

1950

(edited with W.G. East). *The Changing Map of Asia: A Political Geography*. Methuen, London. Pp. 1-29 Introduction: the Asian background; 119-78 India and Pakistan; 385-401 Epilogue: the unity of Asia. Fifth edition 1971 (edited with W.G. East and C.A. Fisher). Methuen, London. Pp. 1-29, 127-91 and 635-48.

(with E. Ahmed) Five cities of the Gangetic Plain: a cross section of Indian cultural history. *Geographical Review* 40: 260-78.

Taking stock in India and Pakistan. *Geographical Review* 40: 279-84.

On the marchlands of geography. *Geographer* 2(2): 1-5.

Contributions [annual editions] on India and associated regions. *Chambers's Encyclopaedia* (New Edition). Newnes, London (to 1965) and thence (New Revised Edition), Pergamon Press, Oxford.

Contributions [annual editions] on India and associated regions. *Collier's Encyclopedia with Bibliography and Index*. Crowell-Collier, New York.

1951

Land, water and man in India. Pp. 7-10 in *Circular to Schools No. 64*. University of Melbourne, Melbourne.

More myths about Asia. *Eastern World* 5(1): 37.

Can India's millions be fed? *Listener* 12 April: 567, 583.

Geomorphological problems in India: a plea for their more intensive study. Pp. 181-5 in *The Indian Geographical Journal Silver Jubilee Souvenir Volume and N. Subrahmanyam Memorial Volume*.

1952

A suggested regional division of the Indian sub-continent. *Indian Geographical Journal* 26: 1-14.

(with a contribution by C.D. Deshpande) The Indian village. *Geography 37:* 142-52.

Toynbee and Huntington: a study in determinism. *Geographical Journal* 118: 406-24.

1953

The Compass of Geography. Inaugural lecture, The Australian National University. Canberra. Pp. 30.

Changing native agriculture in New Guinea. *Geographical Review* 43: 151-72.

Reflections on Toynbee's "A Study of History": a geographer's view. *Historical Studies Australia and New Zealand* 5: 324-37. Reprinted 1956, pp. 287-304 in M.F.A. Montagu (ed.), *Toynbee and History: Critical Essays and Reviews,* Porter Sargent, Boston, Mass.

1954

One-sixth of the world. *Australian Geographer* 6(3): 41-4.

Social structure and function. Pp. 228-40 in H.L. White (ed.), *Canberra A Nation's Capital*. Angus and Robertson, Sydney.

India and Pakistan: A General and Regional Geography (with a chapter on Ceylon by B.H. Farmer). Methuen, London; Dutton, New York. Pp. xxxvi + 827. Translated 1957 into Russian as *Indiya i Pakistan: obshchaya i regionalnaya geografiya*. Izd-vo inostrannoi literatury, Moscow. Pp. 834. Third edition 1967 (revised with A.T.A. Learmonth), Methuen, London.

Reflexões sobre a epopeia Portuguesa no Oriente. *Sociedade da Geografia de Lisboa Boletim* 72: 65-76.

The rationale of modern geography. *Gazette of the University of Western Australia* 4(1): 1-4.

The Pacific: some strategic considerations. *Special Problems of the Member Nations: Australia*. (Australian Papers, Commonwealth Relations Conference, 1954.) Australian Institute of International Affairs, Melbourne. Pp. 18.

1956

Australia New Zealand and the Pacific. Oxford University Press, Oxford. Pp. 96. Third edition, 1969.

The nature of political geography. *Australian Geographer* 6(5): 29-31.

Problems of development in New Guinea. *Geographical Journal* 122: 430-6.

Two federal capitals: New Delhi and Canberra. *Geographical Outlook* 1: 1-8.

Modern geography and its meaning for India. *National Geographical Journal of India* 2: 175-85.

Bush and city: some reflections on the Australian cultural landscape. Presidential Address, Section P, ANZAAS 1955. *Australian Journal of Science* 18: 177-84.

The resurgence of Asia, pp. 451-75; India and Pakistan: internal political geography, pp. 476-504; The neighbours of India and Pakistan, pp. 505-18; The Pacific: some strategic considerations, pp. 519-33; and Australia and its dependencies, pp. 803-30 in W.G. East and A.E. Moodie (eds.), *The Changing World*. Harrap, London.

Monuments of imperial rule in India. *Listener* 26 July: 122-3.

1957

India's economic plan. *Current Affairs Bulletin 21:* 19-32.

Konarak and Sindri: fertility ancient and modern. *Meanjin Quarterly* 16: 341-53.

Westward the course: impressions of Portugal and Brazil. *Quadrant* 2(5): 69-74.
Manuel Godinho de Erédia: quest for Australia. *Meanjin Quarterly* 16: 109-22.
Terra Australis — Cognita? *Historical Studies Australia and New Zealand* 8: 1-19.
How determined is possibilism? *Geographical Studies* 4: 2-12.
'Region' as a term of art. *Orbis* 1: 343-51.

1958

Aspects of the city in South Asia. *Confluence* 7: 16-28.
The end of an old song? The determinism-possibilism problem. *Geographical Review* 48: 280-2.

1959

The 'Hydraulic Society': K.A. Wittfogel's "Oriental Despotism: a Comparative Study of Total Power" (review article). *Annals of the Association of American Geographers* 49: 90-5.
The Fijian People: Economic Problems and Prospects. Legislative Council of Fiji, Paper No. 13. Government Printer, Suva. Pp. 112. Translated into Fijian as *Na Kawa I Taukei: Na Veika Vereverea Vakailavo Me Baleti Ira Kei Na Veika Me Caka Kina.*
Geography and racism. Pp. 124-33 in S.M.T. Rizvi and M. Raza (eds.), *Proceedings of the International Geography Seminar Aligarh Muslim University.* Department of Geography, Aligarh.

1960

'Integrated Surveys' in Australia. *Oriental Geographer* 4: 59-61.
The Burns' Commission report. *Journal of the Polynesian Society* 69(1): 54-6.
Fijians "Must Foot Their Share Of The Bill". *Pacific Islands Monthly* 30(9): 33-9.
Time for tapa curtain to be taken down. *Pacific Islands Monthly* 30(10): 53-7.
Under two laws: the Fijian dilemma. *Meanjin Quarterly* 19:166-81.
Lord Kelvin rides again [guest editorial]. *Economic Geography* 36: facing p. 95.
New perspectives in geography. *Australian Journal of Science* 22: 436-9.
Geography in a quantifying age. *Geographical Review* 50: 124-5.
Quantity and quality in geography. *Annals of the Association of American Geographers* 50: 377-94.

1961

Diversity and unity in the sub-continent. Pp. 178-80 in A.R. Desai, *Rural Sociology in India* (Third Edition). Indian Society of Agricultural Economics, Bombay. (Reprint of pp. xxx-xxxii in *India and Pakistan: A General and Regional Geography*, 1954.)
On being an expert. *Quadrant* 5(18): 25-34.

1962

University geography in Australia. *Australian Geographical Record* No. 4: 3-8.
Finis coronat opus? [A review article on Toynbee's *Considerations*] *Australian Outlook* 16: 84-9.

1963

Islands and men. Pp. 253-64 in F. R. Fosberg (ed.), *Man's Place in the Island Ecosystem: A Symposium.* (Tenth Pacific Science Congress, Honolulu, Hawaii, 1961.) Bishop Museum Press, Honolulu.
Theory and practice in Soviet geography. *Australian Geographical Studies* 1: 18-30.

1964

(with G. Currie and J.T. Gunther) *Report of the Commission on Higher Education in Papua and New Guinea.* Department of Territories, Canberra. Pp. vii + 335.

Canberra, A.C.T. *Current Affairs Bulletin* 35: 34-48.

1965

The place we are part of? A review-article on south-east Asia. *Australian Geographer* 9: 303-6.

Let Me Enjoy: Essays, Partly Geographical. Australian National University Press, Canberra. Pp. xii + 303. [1966, Methuen, London.]

A territory university: cheap at £5,343,000. *New Guinea and Australia, The Pacific and South-east Asia* 1(1): 23-7.

De Lozier Bouvet and mercantilist expansion in the Pacific in 1740. Pp. 221-37 in J. Parker (ed.), *Merchants and Scholars: Essays in the History of Exploration and Trade Collected in Memory of James Ford Bell...* University of Minnesota Press, Minneapolis.

(pseudonym Oliver Skelton) Kitchen stew. *Nation* No. 174: 13.

1966

Education and its problems. Pp. 117-34 in E.K. Fisk (ed.), *New Guinea on the Threshold: Aspects of Social, Political, and Economic Development.* Australian National University Press, Canberra.

Between Tasman and Cook: Bouvet's place in the history of exploration. Pp. 174-86 in J. Andrews (ed.), *Frontiers and Men: A Volume in Memory of Griffith Taylor (1880-1963).* Cheshire, Melbourne.

1967

The last voyage of James Cook. *Australian Journal of Science* 30: 184-6.

H. and M. Sprout "The Ecological Perspective on Human Affairs. With Special Reference to International Politics" (review article). *History and Theory* 6: 464-8.

1968

Australia. Benn, London. Pp. 328.

Environmentalism. Pp. 93-7 (vol. 5) in D.L. Sills (ed.), *International Encyclopedia of the Social Sciences.* Macmillan and The Free Press, New York.

The muse of mercantilism: Jago, Grainger, and Dyer. Pp. 119-31 in R. F. Brissenden (ed.), *Studies in the Eighteenth Century: Papers Presented at the David Nichol Smith Memorial Seminar Canberra 1966.* Australian National University Press, Canberra.

Mandalay and Rangoon: the old and the new in Burma. Pp. 155-68 in *Land Use and Resources: Studies in Applied Geography. A Memorial Volume to Sir Dudley Stamp.* Institute of British Geographers. Special Publication No. 1.

Meanwhile, at June Valley . . . students, strains and some success. *New Guinea and Australia, The Pacific and South-east Asia* 3(4): 32-7.

1969

Britannia, Anglia, Melanesia. Pp. 661-71 in *The History of Melanesia.* (Papers read at Second Waigani Seminar, 1968.) Research School of Pacific Studies, The Australian National University, Canberra.

Western humanism and eastern cultures. Pp. 191-200 in I. Edwards (ed.), *A Humanist View.* Angus and Robertson, Sydney.

Foreword. Pp. ix-xii in G. A. Wood, *The Discovery of Australia* (revised by J. C. Beaglehole). Macmillan of Australia, Melbourne.

1970

The nature of Australian exploration. Introductory chapter, pp. 4-32 in E. H. J. and G. E. E. Feeken, *The Discovery and Exploration of Australia*. Nelson, Melbourne.

Foreword. Pp. vii-x in E. K. Fisk, *The Political Economy of Independent Fiji*. Australian National University Press, Canberra.

1971

Taking stock after twenty-five years. *The Australian National University News* 6(2): 1-4.

Melanesia: an island world. A review of H. C. Brookfield, with D. Hart, "Melanesia: a geographical interpretation of an island world". *Geographical Journal* 137: 557-9.

An epistle to an editor. *Newsletter* Australasian and Pacific Society for Eighteenth Century Studies 1: 6-7.

1972

Models and options for Nuiguini: some informal reflections. Pp. 3-11 in M. W. Ward (ed.), *Politics of Melanesia*. (Papers read at Fourth Waigani Seminar.) Research School of Pacific Studies, The Australian National University, Canberra.

Background to Bangla Desh. *Quadrant* 16(75): 9-14.

(with J. N. Jennings) Australian geography 1951-1971. *Australian Geographical Studies* 10: 113-40.

The last of lands. Pp. 67-78 in G. C. Bolton (ed.), *Everyman in Australia: Octagon Lectures 1970*, University of Western Australia Press, Perth.

Problems and priorities: summing up the sixth Waigani. *New Guinea and Australia, The Pacific and South-east Asia* 7(2): 50-62.

Journeyman Taylor: some aspects of his work. *Australian Geographer* 12: 115-22.

Guest editorial on research in the Pacific. *Pacific Islands Monthly* 43(10): 5.

(with A. T. A. Learmonth and collaboration of A. M. Learmonth) *India and Pakistan: Land, People and Economy*. Methuen, London.

(with A.T.A. Learmonth and B.H. Farmer and collaboration of A. M. Learmonth) *India, Pakistan and Ceylon: The Regions*. Methuen, London.

On atrocities. *Sonar Bangla (Golden Bengal)* No. 6: 6-7.

1973

Foreword. Pp. v-vii in H.C. Brookfield (ed.), *The Pacific in Transition: Geographical Perspectives on Adaptation and Change*. Arnold, London.

1974

Ellsworth Huntington: a geographical giant. *Geographical Journal* 140. 117-19.

Âmes damnées: Baudin and Péron. *Overland* 58: 52.7.

1975

Editor of J.W. Davidson, *Peter Dillon of Vanikoro: Chevalier of The South Seas*. Oxford University Press, Melbourne.

1976

Finis coronat opus: Beaglehole's life of Cook. *Journal of Pacific History* 11: 247-52.

Canberra: the Commonwealth's company town. Pp. 422-39 in J.V. Ferreira and S.S. Jha, *The Outlook Tower: Essays on Urbanization in Memory of Patrick Geddes*. Popular Prakashan Private Ltd, Bombay.

1977

Prolegomena to a history of the Pacific. *Geographica Polonica* 36: 217-23.

'South Sea' to 'Pacific Ocean': a note on nomenclature. *Journal of Pacific History* 12: 205-11.

Canberra. Pp. 452-62 (vol. 1) in *The Australian Encyclopedia*. Grolier Society of Australia, Sydney.

Plus Ça Change . . .? Some Problems of the Sixteenth and the Twentieth Centuries. Annual lecture, The Academy of the Social Sciences in Australia, Canberra. Pp. 29.

1978

The Pacific as an artefact. Pp. 32-45 in N. Gunson (ed.) *The Changing Pacific: Essays in Honour of H. E. Maude*, Oxford University Press, Melbourne.

Palaeoclimates of geographical thought. *Australian Geographer* 14: 1-7.

Sir Archibald Grenfell Price [obituary]. *Geographical Journal* 144: 180-1.

Who came first? K. G. McIntyre's "The Secret Discovery of Australia". *Australian Book Review* 1: 21-2.

Two trials in Burma. *Hemisphere* 22(11): 8-10.

1979

'Baudin and Flinders'. Pp. 87-93 in R.W. Russell (ed.) *Matthew Flinders: The 'Ifs' of History*. The Flinders University of South Australia, Bedford Park.

The Pacific Since Magellan: I. The Spanish Lake. Australian National University Press, Canberra. Pp. xxiii + 353.

Introduction

J. N. Jennings and G. J. R. Linge

Geographers have not been as ready to consider personal action in changing the face of the earth as historians have in weighing the role of the individual in moulding the temporal course of human affairs. Oskar Spate, however, has much of the historian in him so he shares that interest in the individual; Erédia, Bouvet and others have attracted his pen which does not avoid the quirks of character behind a man's mark in the world. This direction is to some extent followed by each of the first four chapters in this book.

The earliest, fruitless efforts at agriculture in northern Australia by the British (Chapter 1) were small so, in accounting for their checkered stories, the protagonists have to take the stage and the part played by human fallibility looms large even beside the acerbities and unreliability of nature in the North.

Personal diaries and letters are the means whereby in Chapter 2 the experience and anticipations of early settlers in their travel at a slow pace half-way round the world are discerned; their immediate reactions to the reality of Australia were often harshly different from their imaginings of the land to which they were entrusting themselves.

The third chapter shifts to later times when a growing national consciousness distinguished the 'pom' and public backbiting already rivalled betting on horse races as the national pastime. The three British transport engineers discussed all suffered from these traits of colonial society as well as other vicissitudes besetting an expert in a distant land yet each reshaped the landscape in varying ways.

Next in Chapter 4 the success or otherwise is considered of just one visiting scientist to whom Australia was remote in more than a narrowly geographical sense. Since research is not complete until it reaches those for whom it is relevant, the appraisal is a mixed one, as it is for most human affairs, though the debits are largely the fault of Australians rather than of their guest.

This group of essays in the historical geography of, and the history of geography in, Australia is followed by another set more formally styled in historical geography, yet ranging widely in method. All are located within the region with which the Research School of Pacific Studies is concerned. Spate's earliest contributions to historical geography — on London's growth — were written when the 'cross-sectional' approach was very much the mode but they conform to that pattern only to a degree. One of us recollects being enthralled when walking the streets of London with him, for, as each building was passed, its antecedents one after another were recalled in fascinating detail backwards in time to the beginnings of clear record. His doctoral thesis, unfortunately never published in its entirety, combined both evolutionary unravelling and spatial 'stills'. Variety of method in historical geography will not seem *ultra vires* here.

Chapter 5 is a plea for the geographer to follow the path of some modern historians, armed with tape-recorder, into the streets, byways, pubs and homes, searching out history in the minds of people who are not likely to set it down on paper. The link with the previous group of writings is evident for the case study documented is the mental map of a particular Australian of his native town.

There follows in Chapter 6 a study of Bangkok in 1880, a cross-sectional portrait of a city, dependent in part for its elucidation and verification on preceding and succeeding cross-sections. The source materials (themselves compared) for the 1880 vignette are alien residents' descriptions and a postal directory which included more than might be expected of such a roll. The Asian city was one of Spate's interests and it is apposite to note that in the nineteenth century the Thai capital was moved towards the growing external danger from European expansion in the way of capital cities in accord with the thesis of Vaughan Cornish which Spate later elaborated.

Likewise rural settlement patterns in peasant societies engaged Spate's attentions so the theme of Chapter 7 falls properly in place here. Successive cadastral plans of Fijian sugar plantations reveal how the past constrains the present land ownership mosaic and the same is true of the social and economic relationships behind it.

Chapter 8 encompasses a much wider time span than the preceding contributions. This account of tropical mountain forest in New Guinea draws on the sequences of pollen and charcoal numbers through lake and swamp deposits to which an absolute chronology has been conferred by the boon of radiocarbon. This contribution is a reminder that it was in Spate's Geography Department that Australian Quaternary palynology made its first firm start. The essay's theme, the past effects of New Guinean agriculture on the mountain forests and the counsel they yield for management of forest (and soil) resources today, bears strongly on present day economic development in Papua New Guinea with which Spate has himself been directly concerned.

There is an easy transition from this final essay concerned with time past to the third group of contributions devoted to present day problems of policy (and geography) in some of the Third World countries within the area of interest of the Research School of Pacific Studies. These illustrate the variety of matters on which much of Spate's energies were concentrated during his terms as Head of the Geography Department and later as Director of that Research School.

Change in a Papua New Guinea highland valley during a recent thirteen-year period is recorded and assessed in Chapter 9. One aspect of that change is, in minuscule, the process of encroachment on the rainforest which is weighed in the balance in the previous essay. The main attention is with the effects of distance, which operate exponentially in mountain country: the village at the end of the line of communications lags more and more behind when modernisation is proceeding. The *cri de coeur* of this author is that policy should aim not so much at making the country like the town as making the town like the country.

Chapter 10 examines the growth of a rural proletariat in Papua New Guinea. One conclusion emerges: that their social and economic transformations are not readily to be fitted into concepts and terminologies devised for societies remote in place and style. This recalls that the way that some Marxists have thrust everything into a rigid framework of fixed concepts contributed to Spate's early departure from that political fold even though he did not abandon the position that economic factors, often geographically founded, profoundly affect other facets of human affairs, including the political.

The claims for tourism as a mode of economic growth to be promoted in Third World countries receive critical examination in Chapter 11 with respect to Papua New Guinea and policies towards this form of economic development are adumbrated for the government in relation to its declared national objectives.

Spate advised on developmental policy for both Papua New Guinea and Fiji; with the latter it was a one-man task. His report on Fiji is evaluated in Chapter 12 and the fate of his recommendations for the Fijian people is reviewed in the light of subsequent happenings. That the Fijian author of this essay is a former research student of his Department and is now a senior public servant in his own country needs underlining here.

The final essay on Third World problems (Chapter 13) reviews city planning practice in Indonesia against the background of indigenous and colonial city structure. It is argued that, perhaps without conscious intent, this planning promotes the Westernised and monetised sector at the expense of the traditional and informal sector.

The last two chapters properly bring the book back to Australia, in fact to the empty heart of this land. Australians have held, and still hold, more false notions about the arid core and the North than about anything else relating to this nation. More than once Spate has provided a just view of what has given rise to 'an almost obsessional compound of guilt and fear' but, as he says, 'any Asian who has flown over the red and grey deserts of the centre — and most Asian visitors do so, is unlikely to have any illusions of a land of plenty greedily withheld from them'.

Chapter 14 analyses how the difficulties facing agricultural and pastoral use of the continent's arid lands destroyed preconceptions about the politically desirable manner of settlement and led to much more governmental involvement than was contemplated or thought advisable by most people. Thus, concepts of land management arose which are so allied to the precepts of the contemporary conservation movement that the question arises whether in Australia there is not in part a relationship between the two of progenitor and progeny. If this is so, it lends a respectability to the latter hardly accorded it in many influential quarters of society.

The one activity generally successful, at least in commercial terms, in what Griffith Taylor called 'Empty Australia', is mining. Despite the views of extreme conservationists, this must be allowed to have had consequences, including beneficial ones, other than simply pockmarking the face of Australia with big holes and dangerous shafts. However these isolated mineral resources have given rise from time to time to ideas of basing manufacturing industries on them. The problems found by Whyalla, which Taylor placed just inside the bounds of the complementary 'Economic Australia', might have given pause to this kind of proposition in recent years. The reasons for the rise and fall of the most grandiose and most recent of all such schemes inside 'Empty Australia' — the industrial complex conceived for Pilbara — are set out in Chapter 15. Politicians in Australia and elsewhere might do well to acquaint themselves with Spate's probabilism even if their predecessors did spurn Taylor's determinism.

Strangers in the Land

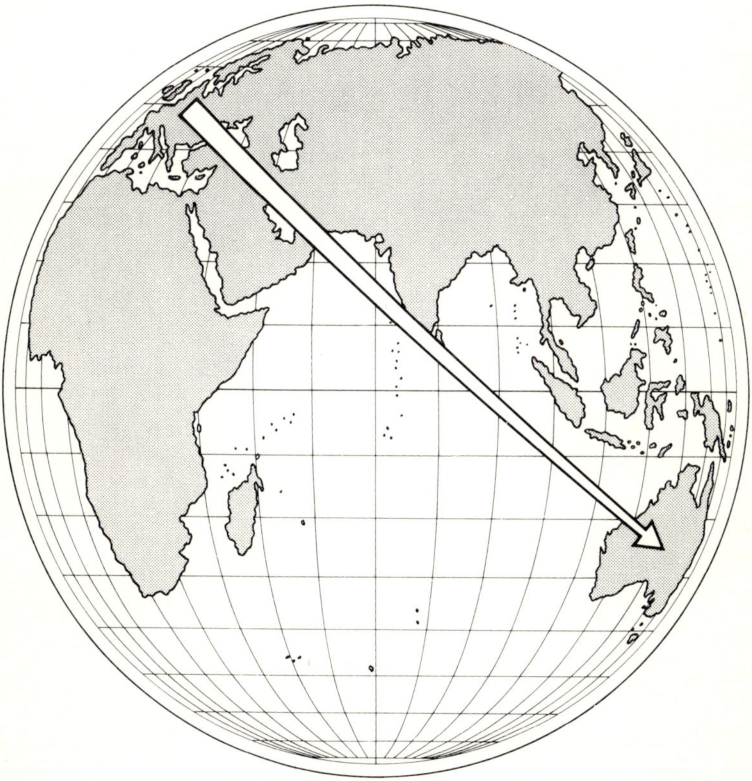

Chapter 1

Trial and Mostly Error

*F. H. Bauer**

On 7 April 1818 Alan Cunningham, botanist and explorer, made the first of what was to become a long line of attempts to induce the soil of North Australia to produce Western domesticated plants. The place was Mullett Bay, on the west side of North Goulburn Island, and the words of Phillip Parker King, captain of the *Mermaid*, 'it is very doubtful whether any have succeeded' (King 1827 I: 72) were prophetic for much more than Cunningham's lonely and soon-abandoned 'garden'.

Shortly afterwards the Colonial Office was debating the advisability of establishing a settlement on the north coast of the continent (Howard 1933: 76). Among the several arguments put forward, the possibilities of profit from products of the soil rated only a throw away paragraph (*Historical Records of Australia [HRA]*, III 5: 745): 'The Climate moreover affords a hope that, at a future period, the articles of Coffee and Tobacco may be cultivated with considerable Success'.

However, strategic and trading arguments were quite sufficient to carry the day, and from these beginnings ultimately came three

* *F.H. Bauer, a graduate of the University of California (Berkeley), was the second Research Scholar and later Research Fellow, in the Department of Geography, Research School of Pacific Studies, The Australian National University. Between periods of lecturing in Californian Universities, he created the Geography Department at the University College of Townsville. He is now the first Director of the North Australia Research Unit, The Australian National University, at Darwin.*

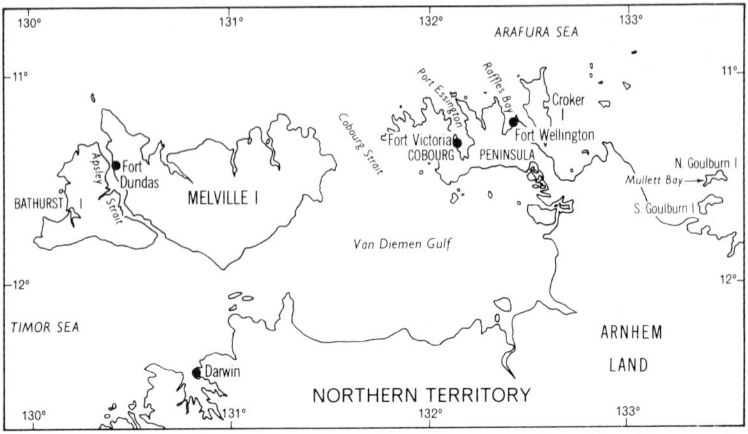

Figure 1.1 Part of northern Australia. Places mentioned in Chapter 1.

essentially military settlements whose garrisons had the doubtful honour of being the first Europeans to cultivate the soil of the Northern Territory. These were Fort Dundas on Melville Island and Forts Wellington and Victoria on Cobourg Peninsula (Fig. 1.1); all engaged in what may best be termed subsistence horticulture, not from any love of the Arcadian life but under orders based in necessity. The 'tyranny of distance' laid heavily upon the settlements, combining with dangerous and imperfectly known sailing and weather conditions, to make regular supply both difficult and expensive; it made sense to grow as much of their own food as possible. Scurvy and other deficiency diseases were prevalent and fresh fruits and vegetables were known to alleviate and cure them. It was also important to discover what would grow and how well, particularly items, such as spices, with a potential for trade. To these several ends seeds and cuttings of a variety of plants were supplied to all the settlements from Sydney and these were supplemented from the islands to the north whenever possible or, in one case, from an earlier settlement (*HRA,* III 5: 777, 803). Again prophetically, the results were but meagrely successful and a pattern now long familiar was soon set: initial enthusiasm and optimism gave way to disillusionment, discouragement and finally to despair.

Fort Dundas (Melville Island)

The process is well illustrated by Fort Dundas, the initial settlement, founded in September 1824. It had been intended to locate it at Port

Essington, but a lack of water there led to its establishment on the eastern side of Apsley Strait at a site to be unmatched for unsuitability for more than four decades. On 11 November, less than six weeks after the site had been chosen, and just before leaving, Captain Bremer wrote to the Colonial Secretary (*HRA*, III 5: 777):

> The Soil of this Island [Melville] appears to be excellent . . . [It] is probably capable of producing most of (if not all) the valuable trees and shrubs of the Eastern Islands; all the plants brought from Sydney flourish luxuriantly, particularly the orange, the lemon, the lime, Banana, Napal and Sugarcane; melons and pumpkins spring up immediately, and the Maize was above ground on the fourth day after it was sown.

Bremer sent his despatches to England in the care of Henry Ennis who, as Bremer's clerk, had full access to their contents. When the reports became available in England in April 1825 they received treatment in two rather different quarters.

John Barrow, Second Secretary to the Admiralty and a staunch proponent of English occupation of the north Australian coast, went into one of the flights of fancy which have often marked European opinion of the North, especially those individuals who have had little or no personal experience of it. He wrote (*HRA*, III 5: 793) on the basis of letters from Bremer and others and of charts:

> there never was so promising a spot in a Naval, Com'l, and Agricultural point of view, as the two Islands of Melville and Bathurst and the intervening Strait. The Soil and climate is so fine that there is not an article either of the Temperate or Torrid zone that might not be raised, even the Spices, the wild Nutmeg flourishing in great vigour.

Ennis (1825) promptly wrote a detailed account of the settlement and its prospects for a popular journal, liberally using Bremer's phraseology and adding bits from his own observations. He exclaimed at the rapidity with which seeds appeared above the ground (as do modern visitors to the North), commenting that 'the plantain, prickly pear and loquets never lost a moment from being transplanted', and a variety of other garden plants did well. Common potatoes were a signal exception, but this, he surmised, was because those planted had never been intended for seed, they had been planted too late in the season, the 'professed gardener from Sydney' had not prepared the soil sufficiently and there was trouble with ants. In spite of all, he was convinced that 'there is little doubt, when the land is cleared to a greater extent, and the place comes to be cultivated . . . all these difficulties will be easily overcome, and . . . the excellency of the soil will produce abundantly all the luxuries and necessaries of life'. Furthermore, paraphrasing Bremer, he considered that some of the low-lying land 'holds out the prospect that valuable rice plantations may be made along that part of the coast'.

In addition Lieutenant J. S. Roe, who had been master's mate with King when Apsley Strait was discovered in 1818 and was also in the founding party, sent his former captain a no less optimistic account of the settlement, together with a detailed chart he had made (Fig. 1.2) (King 1827, II: 238, 242).

Meanwhile, back at the Fort all was not well in a commercial,

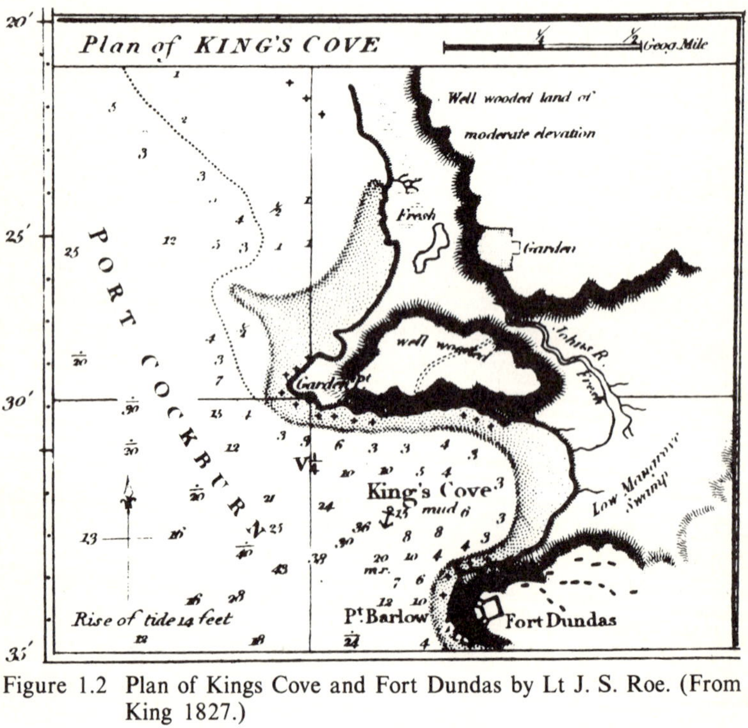

Figure 1.2 Plan of Kings Cove and Fort Dundas by Lt J. S. Roe. (From King 1827.)

agricultural and social sense; no one came to trade, relations with the Aborigines were poor and the general health of the garrison worse. Nevertheless, Captain Barlow, whom Bremer had placed in charge on his departure, maintained confidence: 'upwards of thirty were on the sick list some time ago' and three had died of scurvy, he wrote (*HRA,* III 6: 645):

> I believe the only articles necessary to keep the people in good health are fresh meat and vegetables; of the latter we have every prospect of having a plentiful supply in a short time; both the Soldiers and prisoners huts have gardens attached, in which they have a quantity of water melons, and the pumpkins are ripening fast; when Mr. Miller was at Timor, he made a collection . . . of whatever he thought might be of service to us; to him we are indebted for the Cocoa, Mango, and Tamarind Trees, which are thriving very well; he also brought with him two Yams (all he could procure), which I thought not the least valuable part of the cargo . . . they have all thrown shoots; I will keep the produce for seed . . . the Millet, I brought from Sydney, did very well . . . planted some cotton in January; it is nearly ripe.

The surgeon, Turner, saw it differently; noting (*HRA,* III 6: 650) that they had had no fresh meat and that 'the growth of vegetables has been unproductive'. Barlow thought Turner too 'gloomy' and, while admitting there were not enough vegetables for everyday consumption, claimed they were coming in 'very fast' and should be in good supply for the next six months.

The *Phillip Dundas,* which arrived on 22 February 1826, brought a further supply of seeds and plants from the Sydney Botanical Garden, as well as a prisoner, Richardson, who was to serve as gardener at a wage of £25 per annum. The shipment comprised (*HRA,* III 6: 655):

> Cabbage of sorts; Carrott; Parsnip; Melans of Sorts; Cucumbers of Sorts; Pumkins; Squash; Radishes; Cress of Sorts; Celery; Parsley and Mustard. Plants of Pyne Apples; Bananas; Coffee; Sugar; Cochineal Cactus; Lemons; Oranges; Loquats; Prickly Pear; Thyme; Sago; Marjorum; Peppermint; Spearmint; Tanzy; American Cress; Water Cress; Sarel; Broad French; Hyssop and Cactus for pickling; Capeea and Guinea Grass.

There is no indication whether these species were specifically requested, but probably they were sent simply as being available and possibly useful. Certainly some bore little relation to the environmental conditions they were to encounter. The selection of a number of temperate land plants reflects both the appetites of the garrison and the range available in Sydney. The inclusion of a number of kitchen condiments may have derived from a hope that they would aid in controlling scurvy as well as tease the taste buds. The cochineal cactus, coffee and sugar cane indicate some intent to try plants with an export potential, while the guinea grass may have been an attempt to upgrade livestock fodder.

When Major Campbell (1834: 133) arrived in September 1826 to relieve the founding garrison he found a very depressed group who 'gave us a discouraging account of the oppressiveness of the climate, the scarcity of vegetables, the deficiency of fresh meat, the almost impossibility of procuring fish, the dreariness of the situation . . .'. As befitted a man fresh to a new post he declared himself 'not of a temperament to be cast down by these accounts', but this was to undergo a substantial change. He found fifty-two cleared acres (21 ha), three of which were cultivated, and another ninety-five (38 ha) on which the timber had been felled but not removed.

> The gardens were very backward; there not being any competent person to attend to them until a few weeks previous to my arrival. The soil near the sea was very rocky . . . and on this ground a government garden was marked out in 1824. Melons and pumpkins grew well in it, and where there was any depth of soil it was found to be good, and all seeds planted there sprung up quickly; but what Nature did for them was counteracted by an incapable gardener.

It should be noted that these comments were published, and presumably written, eight years after the events related.

Campbell thought (*HRA,* III 6: 664) the soil 'good at a little distance from the water side', hoped to supply the settlement with vegetables next season, and planned to try 'a little Rice in some of the swampy ground'. Late in 1826 the *Mermaid* was sent to Timor for '20 pekuls of Indian Corn, and 1 pekul of Yams; the latter are not grown at Coepang and the charge for them is high ($5); but, as You wished them for seed, I thought it best to procure them'. The interest the men had in obtaining vegetables was evidenced when they 'gratuitous[ly]' cleared and built a fence around an acre of land for a new garden (*HRA*, III 6: 680).

The wet season of 1826 sorely tried the whole garrison, body and soul; six died, five from malaria. By April Campbell was willing to admit that his first reports might have been a little over optimistic. In this new mood he wrote (*HRA*, III 5: 802-3):

> the baneful influence of the wind is a great drawback upon cultivation . . . [it] breaks down the corn, destroys and roots the fruit trees, and many of the most productive Vegetables. The grass, although it rises to a great height, is not in general nutritious . . . The Soil is in general poor . . .
>
> Our experimental Garden is close to the Sea, and it is only by great labor in removing rocks and introducing a quantity of good soil, that it at all promises to be of any value.
>
> By the assistance of manure and lime, we have lately Succeeded in bringing our fruit trees rapidly forward, particularly the Plantains and Banana . . . but . . . our exertions have been nearly all rendered vain by a very severe gale or tempest on the 2d of April . . . It threw down all our fences, either broke or uprooted mostly every Fruit tree, destroyed a great part of our only present vegetables, the Bringit, completely destroyed all the corn . . .

In May 1827 another shipment of seeds was sent them by Frazer, the Colonial Botanist in Sydney, the list running strongly to brassicas, temperate land root plants, melons and beans.

Campbell's belated but more realistic assessment extended to all aspects of the settlement, including health, supply, navigation and morale, and soon the machinery was set in motion to choose a new site. Still, in retrospect his mood softened somewhat, and in June 1828, after he had been relieved, he noted (*HRA*, III 6: 721):

> I was inabled [*sic*] to make frequent general issues to the whole Settlement "of vegetables". Pumpkins were plentiful, and there was a very promising appearance of yams. Besides the general Crops, the private gardens attached to the huts were presenting gratifying appearances of rewarded industry.

The list he gave in 1834 of forty-two plants tried at Fort Dundas, together with comments on their success or failure, makes it difficult to understand why vegetables should not have been in good supply; only three food crops — potatoes, turnips and peas — were rated as failures.

Major Hartley, Campbell's relief, had no such illusions; not long after taking command he stated (*HRA*, III 6: 716): 'the great sterility and discouraging unproductiveness of the soil throughout Melville Island . . . being for the most part a dry hungry sand, the surface of which is covered with an iron stone of a burnt appearance'. Curiously, there is little discussion of the effects of the dry season on the gardens. Only Campbell (1834: 149) in his retrospective account, comments that 'In exposed places, vegetation is completely burnt up; but, with great care, some garden vegetables are preserved, the chief of which is the pumpkin'. No mention has been found of attempts to water the gardens.

Fort Wellington (Raffles Bay)

The settlement at Raffles Bay was in reality an extension of that at Melville Island, and for a time the two co-existed. It was undertaken primarily to procure a more suitable site, particularly from the commercial standpoint; health was also an important but secondary consideration. Any desire to improve the agricultural possibilities seems not to have been considered.

On 18 June 1827 Captain James Stirling chose the site, took formal possession and named it Fort Wellington. The next day command was turned over to Captain Henry Smyth, livestock were landed and a garden begun. Stirling pronounced the soil 'tolerably good, and equal to any I have seen on this coast', although this contrasted with the opinion expressed by Smyth on his first day in charge (*HRA*, III 6: 769) that it was 'sandy and indifferent'. Smyth was pessimistic even before he got there, and his comments were invariably more critical than those of others. Thus, Stirling recorded (*HRA*, III 5: 815) just before he left on 23 July 1827: 'The Peas are in flower, the radishes nearly fit for pulling, carrots, cabbages, Pumpkins, Bananas, Orange trees and several other things, all well advanced for the short time they have been planted'. Smyth, writing on 17 July (*HRA*, III 6: 771) had been less expansive:

> The Garden with some seeds and plants is promising, in others

> not so; the Peach Trees have taken root but the buds, which were inserted in Sydney, have altogether failed. The Bananas thrive well, also the Sugar Cane . . . The whole of the Ground I have as yet seen, with the exception of the lowest . . . is sterile and composed of a sort of pudding stone, strongly impregnated with Iron . . .

Nine months later Smyth managed to have himself relieved on grounds of ill-health and turned Fort Wellington over to Lieutenant George Sleeman on 24 April 1828.

In September Sleeman noted (*HRA,* III 6: 815) that a durable six-foot fence had been built around the government garden and he was looking forward to a good supply of vegetables. More land had been cleared for maize, sugar cane had done well, as had some pineapples brought from Melville Island. He thought 'the soil . . . capable of producing any thing which the Climate will allow'. Campbell, too, noted (1834: 165) that

> in the second year [1829] . . . the orange, lemon shaddock, cocoa-nut, and tamarind trees were succeeding very well . . . the pine-apple plants were strong and healthy, bananas . . . pumpkins and sweet-potatoes [were abundant] . . . sugar-cane, turmeric, arrow-root, capsicums, and other culinary vegetables, were rapidly increasing . . .

In September Sleeman gave over to 'the first conscientious leadership seen in either of the two settlements' (Allen 1969: 302), that of Captain Collet Barker, who set to work to make a success of the settlement. While his most determined efforts were directed toward bettering relations with the Aborigines, and inducing Malays to come to fish for trepang and to trade, he made a realistic assessment of the site (*HRA*, III 6: 823-4): 'the disadvantages . . . are Chiefly insufficient depth of Water, difficulty of landing, want of water for our Cattle in the dry season, and sterility of Soil . . .'. He surveyed the immediate hinterland to get an idea of the nature of the soil and found it 'generally unfavourable for Cultivation on a large scale . . . [but] I do not despair . . . of cultivating sufficient for our purpose by dint of labour and manure'.

Early in 1829 orders were received to abandon Fort Dundas, and Raffles Bay got a substantial legacy in the form of plants, as well as the gardener, Richardson. It seems clear that Barker was well on the way to making Fort Wellington succeed horticulturally and otherwise. However, Smyth's unrelievedly damning reports reached the Colonial Office well before Barker's, and on 31 July 1829 the settlement was shocked when the brig *Amity* brought orders to withdraw. The *Amity's* surgeon, T. B. Wilson, was clearly impressed with the gardens, noting (1835: 117) that they

> contained orange, lime, and lemon trees, bananas in abundance, shaddocks, citrons, pineapples, figs, custard apples, papaws, tamarinds, dates, cocoa nuts, arrow root, sugar cane, peaches, pumpkins, sweet potatoes, turmeric, capsicum, black pepper, and many other useful and ornamental articles, all of which were thriving well, except the figs and peaches . . .

Wilson and Barker tried to explain to the now friendly Aborigines the uses and virtues of the various plants, but wondered whether any would survive. Barker, who had seriously considered disobeying the orders to abandon the site, clearly had faith in its future; he had Richardson plant bananas, coconuts, custard apples and other trees, and he left poultry and a boar and three sows, a bull and three cows, a horse and a mare in foal for the use of future settlers. Finally, early on 29 August, Wilson noted (1835: 116) that he and Barker 'walked through the deserted camp, and visited the garden, where everything appeared to our eyes more flourishing than before'.

Nonetheless Wilson engaged in some trenchant post-mortem comments. In general the soil could not be called good, being fertile only in patches; agriculture, he felt, had no future because of the high price of labour. Commodore Laws, of the *Satellite*, was even more vehement about gross ignorance of affairs horticultural; if one example is any guide, his condemnation was justified. According to Wilson (1835: 155), Laws asked the settlers whether they had tried any rice; they had, but they had sown *polished* grain; small wonder they got little return.

Fort Victoria (Port Essington)

To rephrase Allen (1969: 306-7), the first two settlements on the north Australian coast were dead but the idea would not lie down. Ever since King discovered Port Essington in 1818 and described its harbour as 'equal, if not superior, to any I ever saw', it had been considered a potential site for a settlement. King was not always prescient, for on that occasion he commented that its location with respect to the islands to the north and the sea route between Sydney and India via Torres Strait ensured that it would soon 'become a place of great trade, and of very considerable importance' (1827 I: 92). Only the fact that Bremer had been unable to find sufficient water there in 1824 had prompted him to move on and finally opt for Apsley Strait, and there was a strong suggestion (*HRA*, III 6: 823) that Fort Wellington was to be occupied only until a shift to Port Essington was ordered. The leaven which had produced the two earlier attempts was still active and it was not long before it again began to rise. All the former reasons were brought forward plus one or two new ones, such as the supposed threat to English sovereignty from France and the United States (Allen 1972: 350). So it was that on 20 October 1838, fourteen years after he had first taken possession of Port Essington, Bremer, now Sir Gordon, again led a founding garrison into that harbour, this time to make the third and last of the Colonial Office's attempts to establish an Australian Singapore. The action, although played out over a longer period than previously, was to have a familiar plot and cast of characters.

Although no specific orders have been sighted, the planners and founders of Fort Victoria clearly expected the garrison to grow a portion of its own food. Before leaving England (Public Record Office [PRO] 1838) the expedition was assigned a gardener, John Armstrong, who was to double as a collector of botanical specimens for the Royal Botanic Gardens at Kew (Spillett 1972: 21).

They came via Rio de Janeiro, where a selection of tropical plants and seeds was obtained (Earl 1846: 165), and while outfitting further in Sydney the Colonial Secretary, Alexander McLeay, saw that they got 'a portion of all that was in the Botanical Garden', including tropical fruit trees and more seeds (PRO 1838). The planting of a garden 'near the sea shore, to the south of the settlement' was among the first tasks undertaken, although Earl, linguist to the settlement, considered the soil to be of a 'light description' (Earl 1846: 165). Bremer was characteristically optimistic. In a personal letter to Gipps he noted, 'The soil all around is as fine as can be imagined' and 'The Marines have built as pretty cottages as an English Village can show', each with its own garden. Indeed, he declared 'as all things are "Couleur de Rose" you may conclude my spirits are high'. To the Secretary of the Admiralty (PRO 1839) he was equally expansive:

> Every day's residence develops the advantages of this splendid position. The soil is exceedingly rich — plantains, bananas, oranges, lemon, and tamarind trees are flourishing. The growth of the sugar cane is astonishing, and cotton succeeds as well as we could wish. The considerable track [*sic*] of meadowland around us are now either flooded or well saturated, and prove that rice may be cultivated in sufficiency for a large population. Our gardens are advancing to perfection; and as I encourage the men to cultivate their own small spots of ground, there has been a sort of rivalry excited, and they produce more than our public and larger garden.

He was quite convinced that rice, spices, cotton 'and almost all tropical Valuable articles . . . may be cultivated with success', although he admitted that there was a shortage of vegetables in the settlement (PRO 1839). He attributed (in May 1839) the lack of success in the public garden to the fact that 'my Botanist is too great a gentleman (In his own opinion) to grow cabbages and pumpkins'. A second garden was established west of the settlement, where the soil and moisture supply were better and exposure to the wind less, and here better results were obtained (NSW Archives Office [NSWAO] 1841).

There is no doubt Bremer did all he could to establish subsistence horticulture at Victoria. He sent the settlement's vessels to Timor and other relatively nearby islands for supplies, and Armstrong went along to obtain suitable plants and seeds (Spillett 1972: 36). Indeed, as long as Bremer was in command, things at Fort Victoria progressed reasonably well, horticulture included. Bremer and his crew stayed until early June 1839 and, when they left, his leadership and the extra hands were sorely missed. Captain John McArthur, who took over as Acting Commandant, was fully convinced of the success of the place and, while admitting his own inexperience in horticultural matters, laboured unceasingly to make the soil yield the largesse of which he believed it capable.

One of his first problems was Armstrong, the gardener, who had become increasingly dissatisfied and who, it appears, knew little of tropical horticulture. Bremer had insisted that Armstrong's major duty was to attend to the gardens and had sharply curtailed his collecting activities, which Armstrong felt were the really important part of his assignment. He complained bitterly of the hopelessness of making Victoria self-sufficient in garden produce, of the treatment he received and of many other injustices (Kew Correspondence, 23 November 1839, 7 December 1840). Probably no one was greatly upset when, in November 1840, he left to live in Timor (Earl 1846: 165).

During its first two years the Victoria garrison was subjected to nearly the full range of climatic hazards the North sends to plague those who would till its soil. On 25 November 1839 the settlement was struck by the first tropical cyclone to be described from this part of the continent. It did a good deal of damage, laid the gardens waste and set everything back six months or more (NSWAO 1840a). The wet season which followed the cyclone was, to judge from McArthur's records, a 'light' one (NSWAO 1842) but the gardens suffered from continuously cloudy weather. Many plants ran strongly to 'exuberant foliage' and rats got nearly as many melons as did the garrison (NSWAO 1840b). By July the other side of the climatic coin was uppermost, and McArthur was likening the dry southeast winds to the 'piercing N.E. [winds] of England . . . in Spring', and noted that all garden produce had suffered accordingly, especially the maize in the old garden which was most exposed (NSWAO 1840c). The wet season of 1840-1 was another trial with weeks of gloomy, overcast conditions; McArthur contended there was not a thirty-six hour break in the rain from 1 December to the end of February, and his records (NSWAO 1842) bear him out. The gardens 'resembled bog lands', deep trenches had to be cut to provide drainage; even so the gardens were much damaged (NSWAO 1841).

In spite of all, McArthur was determinedly optimistic. In February 1840 the gardens were 'objects of . . . gratification' and he considered (NSWAO 1840a) that 'all things being favourable [they would] provide a supply of vegetables in about another [few] months for all the Garrison . . . [and] in future they will yield . . . every fruit common to tropical climates'. Even in the dry blasts of July he was convinced (NSWAO 1840c) 'that whenever we can spare more hands . . . we shall cultivate, with some certainty, sweet potatoes, yams, maize, and such like, sufficiently for general consumption'. A few months later he boasted, 'The Gardens have produced eighteen hundred pounds of vegetables, chiefly pumpkins' as well as some melons, while he hoped for a regular supply of sweet potatoes (Allen 1969: 376).

Only the simplest steps appear to have been taken to mitigate the annual winter droughts. McArthur once noted using two men to water plants and probably this practice was used to some extent every dry season (Allen 1969: 378 quoting Anonymous 1843: 30). Irrigation practices, even simple ones, were not a part of the English cultural heritage, but McArthur seems to have thought about something of the sort, for in one of his earlier despatches he noted (NSWAO 1840d): 'There are situations . . . where water could be reserved on some of the lowest of the second levels; and consequently be made available for the lower level — and this last most generally falls by steps, so that the water is easily controlled'. The Port Essington site did not lend itself to simple irrigation methods and McArthur never had the men or, perhaps, the technological knowledge to put the idea into practice.

McArthur's reports are not the only ones from which a horticultural view of Victoria may be taken; for a place so isolated there were a rather surprising number of visitors. One was Sir Everard Home, captain of the warship *North Star,* who came in 1843 under the orders of the Foreign Office charged with reporting on the state of the settlement (Spillett 1972: 89). Home must have been infected with McArthur's *couleur de rose* mood, for he commended almost everything he saw. He observed the two gardens, each about 1½ acres [0.6 ha], and the productive and well-kept private gardens, patches of 'rich earth', good pineapples and cotton; he even found the sandy soil of the first garden healthy because of its dryness. While the lemons had 'thick rind and [were] without juice', the oranges, natural fruit and the guavas were 'not of a good kind', they all grew 'in the most luxurious manner'. He advocated the introduction of bamboo for building purposes, sago for feeding to pigs and guinea grass for the cattle. Home's ideas on agricultural potential were strongly influenced by his conversations with Earl, who had had a substantial role in the decision to establish at Port Essington and was one of the founding party. As a result Home saw the vision splendid: 'Swampy land [could] be cultivated with rice . . . land not fitted to that Grain produces the finest Cotton'; sugar, cacao, mulberry for silkworms and opuntia cactus for the cochineal insect should be tried. While there was 'plenty of good soil', he felt the climate 'far too hot for Europeans to labor in', and advocated the introduction of Malay labour or, better still, Chinese (NSWAO 1843).

One of McArthur's fondest hopes was that capitalists and speculators would establish plantations on the Cobourg Peninsula. He, and all others to express an opinion, were convinced that non-European labour would be necessary and he often tried to attract Malay and Chinese farmers, sailors and merchants to settle. The government hoped to do so too, but with a lower level of expectation lent by distance and fiscal responsibility; while some official steps were taken, they came to nought. In 1842 McArthur bewailed (NSWAO 1842) the impossibility of properly determining the capabilities of the land until 'a few practical men from the places most noted for . . . Rice, Cotton, Coffee [and] Sugar' (such as Java and Amboyna) came to establish themselves. It was a plea often to be echoed later on by successive Government Residents and Administrators. Even after Home's report the Colonial Office wrote (*HRA*, I 23: 458-9) 'it is sufficiently established that in an Agricultural point of view it cannot be considered as a field open for . . . European labourers' and refused to increase expenditure on the place. Eventually even McArthur acknowledged the impossible dream (NSWAO 1847): 'There seems to be no probability of our establishing any Tenantry here, or, Speculation of any kind'.

Victoria certainly had its champions, but it also had rather more detractors and some of them were placed so that their comments got the attention of government officials and the public. Among them were individuals from the complements of several vessels such as the *Beagle, Fly* and *Rattlesnake* which used the settlement as a base during coastal and scientific surveys. One of these was J. Beete Jukes, naturalist on the *Fly*, which called in at Victoria four times between 1843 and 1845. He was not entirely defamatory; he, too, commended the private gardens, the bananas and pineapples, and commented that yams, sweet potatoes, bread fruit, coffee and rice 'seemed to thrive' (Jukes 1847), but also noted 'as the gardens . . . are not equal to a constant supply of vegetables they [the garrison] live during the greater part of the year pretty much as they would on board ship, on salt provisions, biscuits and rum'. He reserved his harshest judgment for the soil and natural pasture:

> in and around the settlement [the soil] seemed of the poorest

> and most sterile description; indeed, it could hardly be said that anything existed worthy of the name of 'soil' . . . I am at a loss to conceive what any would do with land here, if he even had the whole peninsular given him . . . Rice might . . . be raised in small quantities on the borders of the lagoons; but to raise either rice, sugar, coffee, cotton, or any other valuable tropical produce, for *export,* requires both large tracts of rich soil and an abundant population; two things which do not now exist at Port Essington, nor, I will take upon myself to assert, ever can be formed there . . . Scarcely a blade of grass . . . was to be seen over large tracts . . . within half a mile of the settlement, I do not think enough green stuff of any kind could at this time be found to keep a cow, a horse, or a sheep.

His damning conclusion was: 'I believe it to be utterly worthless as a colony, or as an agricultural or commercial possession, and conceive the only argument for our continuing to hold it must be a political one'. He was, of course, quite right.

For Port Essington, however, the tide had turned. After a lengthy investigation, the Colonial Land and Emigration Commissioners reported to the Foreign Office that only its possible strategic position and use as a coaling station gave it hope of survival. The Foreign Office itself dashed the first and the steamship company the second. On 1 December 1849 Victoria joined Fort Dundas and Fort Wellington in oblivion. McArthur was the only person of the founding party there to the end. It would be instructive to know what he thought of his eleven years and forty days at Port Essington, but no such record seems to have survived (Allen 1969: 344-8; 1972: 357-9).

In Retrospect

It remains to make an assessment of these early attempts to grow domesticated plants in North Australia. It is clear that in all three instances there were strong incentives to grow food (and some other) plants, and a good deal of time, effort and expense was invested in their cultivation. Gardens were started as soon as were the military works and their progress received prominent notice in reports, official and otherwise. The efforts had two major aims: to produce food, making the garrisons more self-sufficient and improving general health and morale, and to experiment with plants thought to have commercial potential; the first was certainly the more important.

A comparatively wide range of plants was tried; on the evidence contained in the various sources available in Darwin the list totals nearly 100 and undoubtedly was somewhat longer. Some, such as peaches, apricots, the temperate lands brassicas and root plants, stood little chance of survival; 'Irish' potatoes, peas and turnips were flat failures: they were probably planted largely for cultural reasons. The settlers also tried a number of tropical and sub-tropical plants of which the British had had knowledge elsewhere: pineapples, sweet potatoes, bananas, yams, bread fruit and coconuts. While their success was mixed both as to times and varieties, it was among these they had their most substantial successes. Rather curiously, there is little mention of pawpaw, a fast-growing plant which should have done well and which must have been known to the British. They were not averse to experimenting with plants from Indonesia and Malaya, and on numerous occasions, particularly at Victoria, made considerable efforts to obtain them. However, they showed most interest in fruits, such as jack-fruit, tamarinds, limes and the like, and only the 'Amboyna pea', accidentally introduced, got mention from the substantial range of garden vegetables which must have been available to them. Experiments with commercial crops such as sugar cane, cotton, rice, indigo and coffee were on a very small scale, were not particularly successful and results did not warrant the efforts spent or the optimism generated.

Viewed by modern standards the Northern Territory's first cultivators were no Abels. Probably at no time did they supply their settlements with fruits and vegetables to a reasonable level, let alone to satiation, although at times they must have grown very weary of sweet potatoes, pumpkin and yams. Such a yardstick is, however, unfair, and judged against the level of knowledge and experience of their time and class they come off much better. The reasons for their mediocre showing are not difficult to establish and some of them still plague the Top End cultivator.

Undoubtedly their greatest handicap was lack of knowledge and this expressed itself in many ways. As natives of cool and humid northwest Europe they did not understand the markedly seasonal climate, especially its wet and dry aspects, and the same cultural background denied them familiarity with ways of mitigating the annual drought so fatal to garden plants. None of the settlements was provided with a gardener who knew anything about tropical plants and this ignorance extended to whole garrisons, for the officers were military men who had no more agrarian background than did the predominantly urban men and the volunteer convicts who made up the bulk of the population. Such experience as they had of tropical agriculture led them to make gross over-estimations of soil capability, and they had the bad luck to be settled on some of the continent's most infertile types. In their time this incapacity may be excused, and they did try to correct their mistakes within the narrow bounds of their circumstances.

The limited life spans of the settlements, particularly Fort Dundas and Fort Wellington, contributed to their lack of success, particularly with tree fruits such as mangoes and coconuts. Shortage of labour to prepare and tend the gardens was a serious handicap, especially at fever-ridden Victoria, where they went virtually untended for weeks. Their failure to make any significant use of the plant resources of the islands to the north was due to ingrained cultural preferences, a factor no less apparent in Darwin of the 1970s.

On the whole these reluctant settlers acquitted themselves quite well. They came to a strange environment and by hard work and meagre experience they built up a modest body of horticultural knowledge which should have been very useful to those who came after them. The real tragedy is that there is very little evidence to show that any one ever heeded their accounts. This, too, is fully a part of the pattern in the North.

Acknowledgment

The tyranny of distance still affects all who live and work in the North, including those who would dig into the past with documentary tools. This essay could not have been written without access to copies of documents collected by P. G. Spillett of Darwin and I gratefully acknowledge his courtesy and assistance.

References

Allen, F.J. 1969. Archaeology, and the history of Port Essington. Unpublished Ph.D. thesis, The Australian National University, Canberra.

Allen, F.J. 1972. Port Essington — a successful limpet port? *Historical Studies Australia and New Zealand* 15: 341-60.

Anonymous 1843. McArthur to Bremer, 3 September 1841. P. 30 in *Copies or Extracts of any Correspondence Relative to the Establishment at Port Essington*. London.

Campbell, J. 1834. Geographical memoir of Melville and Port Essington, on the Coburg Peninsula, Northern Australia; with some observations on the settlements which have been established on the north coast of New Holland. *Journal of the Royal Geographical Society of London* 14: 129-81.

Earl, G.W. 1846. *Enterprise, Discoveries and Adventures in Australia*. Madden and Malcolm, London.

Ennis, H. 1825. Remarks on board His Majesty's ship *Tamar*. *Monthly Magazine* (London) 60 (August-November).

Historical Records of Australia (F. Watson, ed.) *Series III Despatches and Papers Relating to the Settlement of the States*, vols 5 (1922) and 6 (1923). The Library Committee of the Commonwealth Parliament.

Howard, D. 1933. The English activities on the north coast of Australia during the first half of the nineteenth century. *Proceedings Royal Geographical Society of Australasia, South Australian Branch* 33: 21-194.

Jukes, J.B. 1847. *Narrative of the Surveying Voyage of "H.M.S. Fly"* (2 vols). Boone, London.

King, P.P. 1827. *Narrative of a Survey of the Inter-tropical Coasts of Australia* (2 vols). Murray, London.

New South Wales Archives Office 1840a. McArthur to Gipps, 11 February 1840, NSW 4/2509.8,1840.

New South Wales Archives Office 1840b. McArthur to Gipps, 12 May 1840, NSW 41/3282, 24.3.1841.

New South Wales Archives Office 1840c. McArthur to Gipps, 16 July 1840, NSW 41/3283, 24.3.1841.

New South Wales Archives Office 1840d. McArthur to Gipps, 2 November 1840, NSW 4/2545.7,1841.

New South Wales Archives Office 1841. McArthur to Gipps, 25 February 1841, NSW 41/7711, 28.8.1841.

New South Wales Archives Office 1842. McArthur to Colonial Secretary, 1 October 1842, NSW 43/811, 1.2.1843.

New South Wales Archives Office 1843. Home to Parker, 19 April 1843, NSW 43/6153, 22.8.1843.

New South Wales Archives Office 1847. McArthur to Thomson, 14 October 1847, NSW 48/2153, 17.2.1848.

Public Record Office 1838. Bremer to Sec. Admiralty, 16 September 1838, 39/108, 27.8.1839.

Public Record Office 1839. Bremer to Sec. Admiralty, 4 April 1839, 39/108, 27.8.1839.

Spillett, P.G. 1972. *Forsaken Settlement*. Lansdowne Press, Melbourne.

Wilson, T.B. 1835. *Narrative of a Voyage around the World . . .* Sherwood, Gilbert and Piper, London.

Chapter 2

Settlers in Transit

*N. M. Wace**

Of all the human migrations that have taken place within historical time, perhaps none has had a greater impact on notions about the natural world and how it might be used for human advantage than the seaborne dispersals of the West Europeans in post-Columbian times. Historical geographers, including Spate, who have studied the first reactions of Europeans to the newly discovered lands in the Antipodes, have concentrated most of their attention upon the accounts of the explorers themselves. But although the explorers were usually instructed to pay attention to those natural productions which might be useful to later settlers, they were not compelled to view the land as a system of resources from which they themselves would have to wrest a livelihood. The settlers who voyaged to the colonies necessarily had a different viewpoint — most were intending to make a new life for themselves and their families indefinitely in new and unfamiliar places.

During the long voyage from Europe, the settlers had plenty of time on their hands: 'Most never forgot the congestion and the monotony of their three or four or five months at sea, or the idleness of the only long holiday they had in their lives' (Blainey 1966: 158). Diaries kept by settlers voyaging to the colonies in the Antipodes tell

* *Nigel Wace is a Senior Fellow in the Department of Biogeography and Geomorphology, The Australian National University. After graduating in botany at Oxford, he took his doctoral degree at Queen's University, Belfast, and spent nine years in the Department of Geography, The University of Adelaide, before moving to Canberra.*

something of shipboard life, places visited, and first impressions of their new homelands. But such material hardly presents an objective view of a great human experience, if only because those who committed their thoughts to paper are not a representative sample of those who made the journey. Blainey (1966) estimated that 90 per cent of the emigrants who voyaged to Australia during the nineteenth century travelled in the steerage class, and that most of these were illiterate. This account of settlers in transit during the period of free-settler emigration from Europe to the Australasian colonies between about 1825 and 1875 is therefore fragmentary, and based on the diaries and letters of only a small sample of well-to-do voyagers, mostly travelling in the cabin and intermediate classes. Some 230 diaries and letters have been examined, mostly at the Turnbull Library in Wellington (*c.* 90), the National Library of Australia in Canberra (*c.* 80), and smaller numbers in Adelaide, Christchurch, Dunedin, Melbourne, Sydney and Greenwich, England.

The Voyage

At the start of the free settler emigration from Europe to Australia, Cunningham (1827) wrote:

> About eighteen weeks is the average passage from England to Sydney, if the ship proceeds direct, the distance by ship's course being about sixteen thousand miles. Many vessels touch however at the Canaries, the Cape Verds, Cape of Good Hope, or Brazils, to replenish their stock and their water, which both a numerous body of passengers and the consumption of water by live stock tend soon to exhaust.

Whatever the season or the vessel, the outward voyage from Europe to Australia or New Zealand fell into two distinct parts, usually of comparable length but of different character and duration.

At the beginning of the voyage from England, a sailing vessel had to buffet her way into the head winds and swells of the westerlies when in the Western Approaches to the British Isles, and as far as about 35°S in the Atlantic (Fig. 2.1). Horsburgh's sailing directory (1826 edition) advised mariners:

> After leaving the English Channel, steer to pass to the westward of Madeira, at any convenient distance exceeding 7

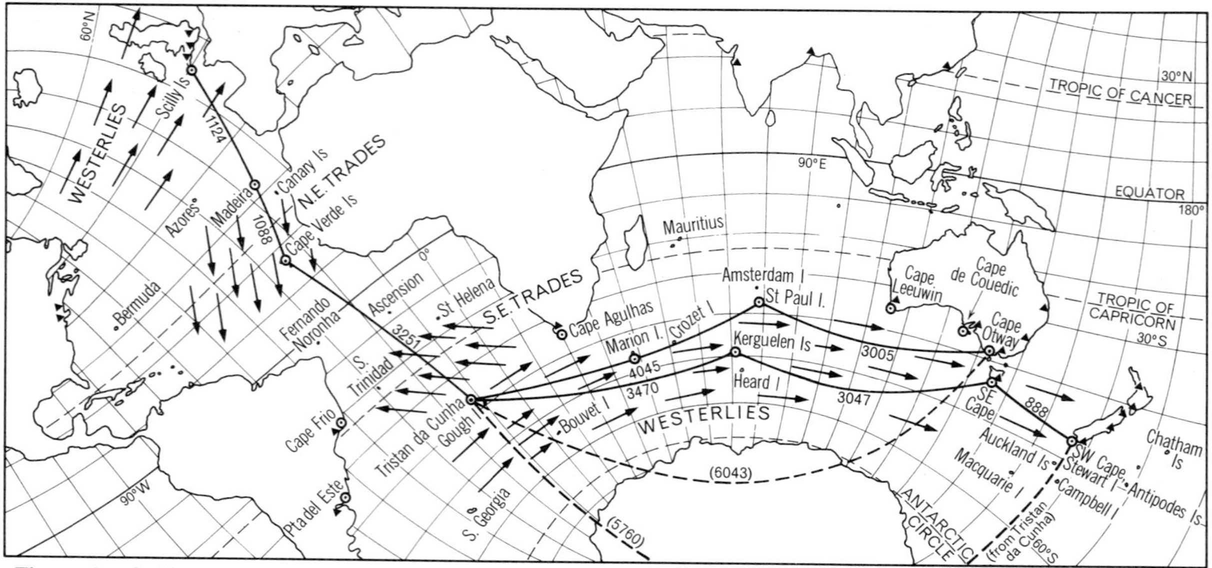

Figure 2.1 Sailing ship routes from England to Australia and New Zealand. Great circles, and the distances along them in nautical miles, between points of departure, some mid-ocean islands, and the usual landfalls on the outward voyages from Europe, are marked in heavy continuous lines. Great circle routes from Tristan da Cunha to Cape Otway, and from Tristan to the Southwest Cape of Stewart Island are marked in heavy dashed lines (that to Stewart Island would go through the South Pole, off the map). Landfalls are indicated by circled dots and ports by solid triangles. Oblique Mercator projection, with its axis passing through position 28°S, 40°E and inclined at 28° to the geographical equator.

> or 8 leagues . . . Departing from Madeira . . . the best tract is to the westward of the Canary and Cape Verd Islands, at any discretional distance, or barely in sight of them.

The islands were sighted in order to check the vessel's position, whether or not stops were made for supplies. After leaving the headwinds of the westerlies between the Azores and the Iberian coast, a vessel had to run through the chancy winds of the 'horse latitudes' before picking up the favourable northeast tradewinds near the Canary Islands, which would take her to 15^{0} or 10^{0}N beyond the Cape Verdes. Horsburgh (1826) advised: 'On whatever side of the Cape Verd Islands ships may pass, the most eligible situation at losing the NE Trade is probably from 18^{0} to 23^{0}W longitude' whence a vessel had to cross the equatorial doldrums of fitful light winds, calm seas and oppressively humid heat of the tropics. The equator was usually crossed between 20^{0} and 30^{0}W, but the duration of the voyage as a whole depended greatly on how long it took the vessel to traverse the doldrums, and how soon and where she entered the southeast tradewinds in the South Atlantic. It had been discovered by Portuguese and Dutch voyagers in previous centuries that the quickest passages to or around the Cape of Good Hope were to be made by sailing across the southeast tradewinds with the wind abeam so as to reach the westerlies in 35^{0} to 38^{0}S near to the South American coast, rather than by beating close-hauled into the tradewinds, and attempting to keep to the eastern side of the South Atlantic.

Before the Australian gold-rushes, this first or Atlantic part of the emigrants' voyages usually took at least two to three months. During this period the emigrants would experience everything from cold, wet, blustery weather with the ship pitching into head seas, to long rolling swells from astern or abeam in the tradewinds, to periods becalmed in the equatorial tropics. Apart from a fleeting glimpse of islands in the North Atlantic, the emigrants might see no land at all. In the South Atlantic, there are no continental ports or islands on the direct sailing route that were easily accessible. Cape Town, which came to be regarded in the years of steamships as the regular half-way port of call, was difficult for outward bound sailing ships to reach directly because to do so they had to beat straight into the southeast tradewinds. The island of St Helena, an important port of call for ships homeward bound to Europe round the Cape, was seldom visited by outward bound vessels for this reason. Emigrant ships therefore tried to take on sufficient supplies for the entire voyage from England to Australia or New Zealand without stopping. Of those which had to stop, but did not make for the Cape, some called at Rio de Janeiro and others attempted to pick up fresh water, vegetables or meat at Tristan da Cunha (37°S) on the fringe of the westerlies. The captain of a vessel calling there in 1821 and 1825 commented:

> Were Tristan da Cunha better known, it might be of great service to ships bound to India, and particularly to New South Wales, as water is very easily obtained, and also livestock and vegetables of every description . . . so that rather than put into the Cape, or Rio Janeiro, it would save time and expense to get refreshment at this place (Herd 1832, quoted in Purdy 1844: 114),

but the lack of a sheltered anchorage and the difficulty of getting ashore through the surf rendered Tristan an unreliable supply point. Emigrant vessels planning to pick up supplies or water on the voyage usually diverted to Rio or Cape Town, although many sighted Tristan to check their positions.

Some of the dilemmas faced by the masters of emigrant ships are well illustrated by the indecisive Captain Hindmarsh of H.M.S. *Buffalo*, sailing from Portsmouth to Adelaide in 1836. Beating into the headwinds in the English Channel at the start of the voyage in July, Hindmarsh was already concerned about fresh water for his passengers and livestock. Off the Isle of Wight, he wrote (Fischer 1962):

> I considered it hardly worthwhile beating backwards and forwards at the back of the Wight, just holding our own, with everybody seasick and moreover, consuming water. The only chance of our being able to pass the Cape [of Good Hope] without calling being to start complete with water; and even then it will require an extraordinary run to enable me to venture to do so.

But by late September *Buffalo* had only reached the vicinity of South Trinidad. Hindmarsh wondered whether he should head for Cape Town, on the direct route to Australia, but into the teeth of the southeast tradewinds; or run with the wind abeam for Rio de Janeiro — an easy sail, but lengthening the distance to be covered. He hesitated, decided to make for the Cape, changed his mind, and finally bore away for Rio, thus lengthening the voyage considerably. The planning and execution of a voyage half-way round the world with a ship full of people and animals were no easy matter before the invention of refrigeration and the desalination of seawater by distillation.

Once into the westerlies in the South Atlantic, at about 35° to 37°S, the whole character of the voyage changed. The rest of the voyage was spent 'running down the easting' in the long rolling swells and following winds of the westerlies. This was the fastest part of the voyage, and if a vessel did not divert to Cape Town (34°S), there were no ports or inhabited islands where supplies might be obtained between Tristan da Cunha and Australia. Without any diversions, a vessel could approach a great circle course from the South Atlantic, but such a route from Tristan to Australia or New Zealand would take her into ice-locked seas, or even to the Antarctic (Fig. 2.1). The routes actually chosen — so-called composite courses — were made up of loxodromes (rhumb lines) often from one island group to another. They were a compromise between the longer distances in the warmer, smoother seas of lower latitudes; and shorter distances in wild rough seas with the danger of encountering icebergs in higher latitudes (Fig. 2.1). Before the arrival of the large, fast and sturdily-rigged American-built clipper ships in the 1850s, mariners were advised to keep to lower latitudes, calling at the volcanic islands of St Paul or Amsterdam (39°S) to check their positions. Horsburgh's sailing directory (1809 edition) advised:

> From December to April, if it is not intended to touch at the Cape, a ship should get into latitude 37° or 38°S about the meridian of London, and keep between 37° and 39° in running down her easting; for the winds will be found as favourable for this purpose in 38° or 39°S, or probably more so, than if she were in a higher latitude . . . If a ship, in the [southern] summer months bound to any part of New Holland, gets into a high southern latitude to run down her easting, some of the islands discovered by the French navigators Marion and Crozet or those first visited by Kerguelen may probably be seen; although it is not advisable to proceed so far southward as to see any of them.

But by 1855 Maury, the American oceanographer, was claiming that a better knowledge of the oceans and their wind systems, collected in his famous *Wind and Current Charts*, was enabling vessels to penetrate safely to 55°S or even higher latitudes, and to reduce the average time taken for a voyage from England to Australia from 124 to 97 days (Maury 1855; Williams 1963). The islands which Horsburgh had advised mariners to avoid were seen frequently, and the distances run were much reduced.

Before the gold-rushes, 'running down the easting' often took six weeks or more, but the emigrants might see no more land than a glimpse of St Paul Island or Amsterdam Island, or a landfall at Cape Leeuwin or the South East Cape of Tasmania before reaching their destinations. In the 1850s and later, the larger and faster clippers might sight Marion Island, the Crozets or Kerguelen, but would usually take a month or so, during much of which the ships would be battened down with decks awash.

At the end of the voyage, the approach to Australia or New Zealand seeking a landfall, usually on a lee shore after a long ocean voyage from Tristan, Cape Agulhas, St Paul Island or Kerguelen, presented the danger of shipwreck. Kangaroo Island on the seaward approaches to Adelaide and the western entrance to Bass Strait were strewn with wrecks. Charlwood (1971) lists and locates twelve wrecks, mostly of emigrant ships, on King Island and around Cape Otway up to 1878, and Chapman (1972) adds several from Kangaroo Island. The very names of the islands south of Stewart Island — 'The Traps' and 'The Snares' — indicate the dangers of the approach to New Zealand. The diaries of emigrants often record not only their excitement at the prospect of sighting their new homelands, but sometimes apprehension that their first sight might also be their last:

> February 1st 1842: No observations for these three days, and except by the dead reckoning we do not know our situation . . . this, with the fog and the appearance of a gale, and the proximity to land, for we must be off the south end [of New Zealand] now, has caused great anxiety to be felt by all on board. There are lots of rocks and an island called the Snares off the south end of Stewart's Island . . . (Fell 1926.)

Although the outward voyage to Australia might take about 125 days at the beginning of the period of free settler immigration, during the 1830s and 1840s passage times varied from 100 to 200 days. The Victorian gold-rushes led to the deployment of American-built clippers carrying emigrants on the Europe-Australia run, starting with the famous *Marco Polo*, a three-decked vessel of 1,625 tons, which sailed from Liverpool to Port Phillip Heads with 930 emigrants in sixty-eight days in 1852 (Lubbock 1948). The clippers were both larger and faster than the British and Indian-built hardwood Blackwall Frigates of less than 1,000 tons, which preceded them; but, being built of softwood, they were less durable. They soon became 'water soaked', and were classified A1 at Lloyds for only seven years (Lubbock 1957).

Shipboard Life

Life aboard the emigrant ships sailing from England to Australia and New Zealand was extremely crowded for those who travelled steerage, and monotonous for all. Day after day of routine, in which the weather, the occasional ship in an otherwise empty sea, and birds and other creatures seen, provided the only outside variety. It is hardly surprising that the diaries of the emigrants dwell mainly on the social life aboard the ships — animosities and friendships, entertainment, and the social pretensions of fellow passengers. Births, deaths and marriages provide another recurrent theme of comment, as well as the livestock aboard and the food provided.

In the steerage, the emigrants shared a lower deck compartment amidships (Fig. 2.2). Two tiers of bunks were usually arranged along the ship's sides, each boxed in to form a compartment of six feet square and designed to accommodate four adults. Fixed tables and

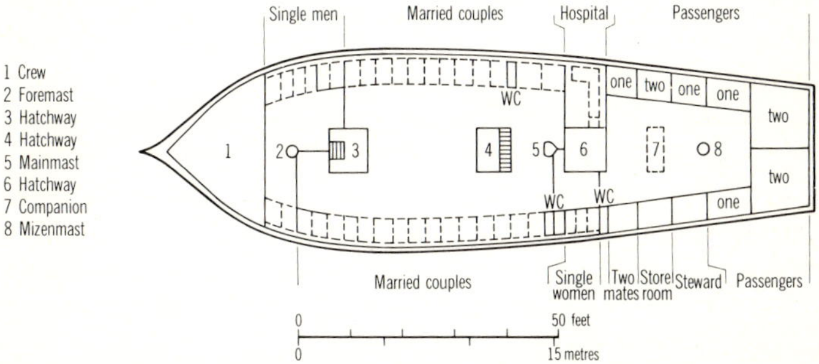

Figure 2.2 Plan of accommodation aboard the *Lord Auckland*. (Based on Alfred Fell 1926.) By courtesy of Capper Press.

Figure 2.3 Cabin passengers' dance aboard the *Randolph* sailing to Lyttelton. (*Source: Illustrated London News* 3 August 1850.) National Library of Australia.

Figure 2.4 Steerage emigrants at dinner aboard an un-named vessel. (*Source: Illustrated London News* 13 April 1844.) National Library of Australia.

benches ran down the midship line, and there was little open space below or on deck (Fig. 2.4). This basic design, the minimum standards for which were rigidly enforced by the Passenger Acts of the British Parliament, probably persisted for most of the period of free settler emigration under sail. Cabin passengers were accommodated aft, as in the *Lord Auckland* (Fig. 2.2). Some vessels carried intermediate class passengers in which families shared more or less enclosed accommodation but messed in the cabin, and some one-class vessels evidently had no differentiation of accommodation.

George Bennett, a cabin passenger aboard the *Lady Clark* which sailed from Plymouth in September 1841 bound for Sydney, described the accommodation and the 200 steerage passengers aboard:

> The ship is divided into three compartments forward for the single men — about 60 — midships for the married and children aft for the single women — about 40. They are mostly Irish, and of the worst description, not six of them are mechanics, and the women are three fourths of them no better than they should be — and all are so dirty, so filthy, that were I to write it you would scarcely credit the account. What the scene and the scent must have been below when all were sick I leave to your imagination 'tis bad enough even in one's own cabin, and I had as comfortable a one as could be got — the aft poop cabin, for which I paid £100 — to include 5 tons of baggage and my dog.

Cabin passengers fared well, at least in some vessels (Fig. 2.3). Barnicoat, voyaging from Gravesend to Nelson aboard the *Lord Auckland*, was apprehensive that the failure of the ship to stop for fresh provisions at Tristan da Cunha would affect their diet. A week after passing the island in December 1841 he wrote:

> I do not know whether we may not have to subsist on salt meat a little, bye and bye . . . the table has always been furnished with fresh provisions so far, but if the rest of the voyage is not pretty quickly performed, our stock can scarcely last. It consists at present of 6 sheep, 3 pigs, 18 fowls, 2 geese and a duck or two. Besides these livestock there are numbers of tins of preserved fresh beef and tripe, soup, also hams, tongues, rice, potatoes, salmon etc. . . . We have fruit tart every day . . . Green (preserved) gooseberries, which are just as fresh as if newly gathered, and red and black currants, cherries, all in their natural state, preserved without sugar, merely by being kept from the air. The potatoes are very good, and the table is kept plentifully supplied with them every day. Plain boiled rice is often set before us . . . Every Sunday we are treated with fresh salmon. A dessert, too, twice a week — sherry and port are drunk every day, and champagne on Sundays in addition. The dessert and tart may be luxuries on shore, but I feel certain that here our uniform good health may in a great measure be attributed to them.

Even in the steerage, passengers got beef or pork every day, as well as bread, tea, sugar and vegetables. If this were the normal scale of rations, they must have seemed luxury indeed to Irish emigrants fleeing from the potato famine after 1840. Fresh water aboard ship is frequently commented on in emigrants' diaries. Fell (1926) recorded that it 'stinks so that to drink any is out of the question', and by the end of his voyage in the *Lord Auckland* during 1842 he was advising future emigrants to bring their own bottled water and a filter. Mary Thomas mentioned a fresh water allowance of 1½ gallons (7 litres) a day aboard the *Africaine* in 1836, occasionally reduced drastically to only one pint (Hope 1968: 51). As in the navy at the time, alcohol (usually in the form of spirits) filled an important function in disguising the taste, if not sterilising, drinking water. A preoccupation with the biological and social effects of intoxication by alcohol has overshadowed its real advantages as a julep aboard ship. The observation by the cabin passengers of its harmful effects in the steerage (e.g. Gouger 1836, in Hope 1968: 47) may have contributed to the growth of temperance attitudes, if not of wowserism, in Australian genteel society later.

A striking feature of emigrants' diaries is the frequent mention of quarrels and disputes. Displays of hostility are obviously over-reported, and many emigrants found in a diary a useful safety valve for personal feelings about their fellow passengers. Ricou (1872), one of the few steerage emigrants to leave a diary, had some revealing things to say about life in the *Indus* voyaging from London to Brisbane:

> a general fight between the Cornish and the Irish . . . there is an ill feeling between the Irish and Englishmen, the former wanting to make themselves too officious. There being the majority of English, the Irish are obliged to be very quiet. Several of both sides were locked up this afternoon.

A few days later he noted: 'another fight in our cabin; there being 110 of us single men, it is usual to see a fight every day'. Before 1851, some 43 per cent of assisted passage immigrants and 33 per cent unassisted passage immigrants were Irish, and 13 per cent and 15 per cent, respectively, were Scots (Borrie 1955). 'Mateship' and group solidarity which are held to be such an important characteristic of Australians (e.g. Ward 1965) would thus have been impressed upon many immigrants from the start of their voyages. Disputes about emigrants' rights aboard ship are common, and are prominent in the accounts of Bent (1814) aboard the *Broxbornebury* and Manning (1839-40) aboard the *Earl Grey*, both voyaging to Australia. Bennett (1841) in the *Lady Clark* describes a series of disputes between the steerage emigrants and the ship's officers, leading on one occasion to fights and near mutiny with a partitioning of the ship, and the issue of pistols and cutlasses to officers and cabin passengers on the poop.

Many emigrants record in their diaries all sorts of organised attempts to keep away boredom: some admitting to the use of a diary itself for this purpose. Isaac Featherston (1841), writing home after arriving at Wellington, commented tersely:

> Suffice therefore now to say that our voyage was like most other voyages, producing nothing of sufficient importance to justify one in wasting a sheet of paper — although most of our cabin passengers dirtied not a few reams in recording the mighty transactions of the voyage, transactions which like most other travels . . . were inventions of their own fertile brains . . .

Less taciturn personalities filled their journals with all sorts of mental bric-à-brac. Ships' newspapers were put together, such as the *Blue Jacket* journal (Anonymous 1855) containing a fantastic collection of stories, poems, sketches and notes 'in a campaign against that formidable foe, "ennui" '. C. A. Brown (1841), aboard the *Oriental* voyaging to Wellington, lamented the 'nakedness of mind' of the other passengers. Determined efforts were made to overcome boredom by clothing such mental nakedness: debates, discussions, navigation classes and formal school teaching were all organised by the cabin passengers. Mary Thomas in the *Africaine* (1836) described one such attempt:

> Seeing that the ill-humour was produced by idleness or ennui it occurred to me that it would be well to get up some general amusement, and I consequently proposed to enrol a body of volunteers to be drilled . . . It is true, sometimes we do not keep very good step in consequence of the motion of the ship, and sometimes a lurch in marching at ordinary time, causes a double quick movement to the rear; but this is all accounted a good joke, and thus the chief end is attained (Hope 1968).

Ricou in the barque *Indus* records a number of activities amongst the steerage passengers during the first month of the voyage from London to Brisbane in 1872:

> *Fifth day*: During this day a great deal of preaching and hymn singing was gone through. It has been quite a treat to see everybody dressed in the best Sunday suit of black cloth just as if we had been at home. *Sixth day*: All sorts of game are carried on cards, dice, dominoes, draughts, chest [*sic*]. Others amuse themselves singing songs and hymns . . . *Eleventh day*: as usual we have a little singing before turning in . . . *Fifteenth day*: School for children opened. *Twentieth day*: The evening closes with a dance open to all, to the airs of Mr Geary's violin, our 2nd Officer and a capital player. As the single girls are not allowed even to look at the single men, [we] were all obliged to take men for partners.

These activities are all mentioned before the fights already described, and thus do not seem to have been altogether successful in preventing discord.

Many emigrant ships carried the livestock needed to establish rural industries in the new colonies as well as to eat during the voyage. To emigrants from rural areas, domestic animals were almost the only point of familiarity in an otherwise alien world, and the behaviour and habits of livestock are frequently mentioned. The *Lord Auckland*, with 170 emigrants, set sail from Gravesend in September 1841 with eighteen pigs, eighteen sheep and some poultry (Fell 1926). The last pig was killed south of Tasmania when four months out, on the five month voyage to New Zealand. The *Cashmere*, bound from London to Auckland in 1851, sailed with thirty dozen fowls, twelve sheep, thirty pigs and a cow for milk. One sheep and three pigs were killed every fortnight, thus allowing for a voyage of at least six months. Carrying such a menagerie had its problems: during the first week of the voyage, one-third of the overcrowded hens died 'much to the relief of the others' (Martin 1851). Pigs were easily fed on scraps from the galley and were commonly carried as a source of fresh meat for the cabin passengers.

On the long voyage, visits to ports or islands, encounters with other ships, and observations of marine wildlife, were subjects of frequent comment in emigrants' diaries. The mere sighting of islands was a matter of eager anticipation and excited interest. Patricia Torlesse (1856), a young cabin passenger in the ship *Egmont* off Tristan da Cunha, wrote: 'All the young ladies (ie. unmarried) got up at ¼ past 4 to see the sun rise upon this island, that blessed island which serves to give us a sight and a feeling that this world was not all water . . .'. South Trinidad in the Atlantic and St Paul Island in the Indian Ocean are described often in emigrants' diaries, usually with a mixture of apprehension and awe, and keen disappointment expressed if they were missed, or passed at night. Passengers were seldom allowed to land at such islands because of the difficulty of re-embarkation should the weather deteriorate; but if cabin passengers were allowed ashore, this often precipitated fierce squabbles about who should enjoy the privilege.

Encounters with other ships gave an opportunity for the despatch of letters, but meetings with those sailing to Europe were rare both because outward and homeward bound vessels followed different routes for most of their voyages, and because captains were generally reluctant to stop merely to exchange mail, news and pleasantries. The period of free emigration under sail included the heyday of American deep-sea whaling, and many encounters were recorded with the ubiquitous Yankees. Fell (1926) related how he and a few other cabin passengers went aboard the whaler *Daniel Webster* in January 1841 and were presented with boxes of cigars 'which were such rascally bad ones that we could not smoke them'. However, the encounter 'advanced the character of the American people much in our estimation', an opinion reinforced by meeting another hospitable Yankee whaler, the *Xenophon* of New York, a few days later. But not all Americans were so hospitable. Taylor (1841), aboard the *Sir John Falstaff* voyaging from London to Sydney, reported: 'A ship passed us this morning, but though we bore up to her, would not alter her course, when 100 yards would have enable us to speak. Probably a damn unaccommodating Yankee, who only stop when it serves their turn'.

Sightings of whales and other marine mammals take a prominent place in emigrants' diaries, but seabirds were apparently regarded as much as targets for pot-shots as creatures to admire. Featherston (1841) commented tersely: 'Our only amusement consisted in shooting the birds which flew about the ship during the whole voyage; but for my part I found ample employment in managing the hundred emigrants. . .'. Such sporting activity aroused the critical comments of some passengers. Barnicoat (1841), two days off Tristan da Cunha aboard the *Lord Auckland*, wrote:

> several of the larger birds (albatrosses) have been shot within a day or two. It is astonishing to see the delight which is displayed at having broken the wing of one. Even breaking a leg is a source of considerable joy. Of course, the after agonies of the poor thing are quite forgotten.

First Impressions of the Colonies

Emigrants' diaries written at sea have little to say about their destinations. Perhaps their expectations would have been based on only limited information consisting mostly of handbooks and guides (Heathcote 1976: 35). But if the emigrants' anticipations of the colonies are meagre, their first impressions of the landscape and its towns and people written when they arrived are often vivid and revealing.

At the end of their voyage, the emigrants were eager for the first sight of their new home, but this was soon tempered by discoveries — sometimes even before they got ashore — that life would not be easy. Lyle (1852), aboard the *Nepaul*, remarked that all the passengers were up on the poop before daybreak as the vessel neared the entrance to Bass Strait at the end of a stormy voyage: 'to behold the promised land. We sailed along at some little distance from the shore, and for the first time beheld the rising of an Australian sun coming up in all its splendour, dispelling the gloom . . .'. As the sun rose higher and became warmer, the decks were covered with flies; as they approached the heads to get into Port Phillip Bay, the sailors refused to work the ship 'as they wanted the ship to go on shore, that they might make their escape' to the goldfields. Passengers had to work the sails to save them from shipwreck, and at Williamstown they anchored 1½ miles (2.4 km) offshore to discourage the crew from swimming to freedom. The ship's boats were chained and locked up overnight. These emigrants were thus introduced early to two features of Australian life — flies and strikes — even before they got ashore. Lubbock (1948) states that by June 1852, the year after the discovery of gold at Ballarat, some fifty vessels were lying in Hobsons Bay off Melbourne, deserted by their crews.

Perhaps inevitably, during the long monotonous time at sea, the emigrants' expectations of their new homes would rise. They had looked forward to release from the confines of a crowded ship, to the freedom of more space, of fresh water and food, of trees and grass and a changing landscape. They would have been less than human if they had not imagined a more beautiful, productive and attractive place than the new land was likely to afford. Not surprisingly many record disappointment among their first impressions, usually vented on the unkempt raw aspect of the new settlements, or on the harsh scrubs.

Isaac Featherston, arriving to take up his post as the Government Surveyor in June 1841, recorded the first impressions of his fellow passengers:

> But where was the renowned city of Wellington? Did these mud hovels scattered along the beach, or these wooden huts which appeared every here and there — did these represent the City of Wellington? Yes. This was the City of Wellington! Where then is the fine fertile land which shall produce such astounding crops? Surely not those steep and wooded mountains? Oh no! These mountains are part of the City of Wellington.

Less well known than his namesake, who wrote the notable *Wanderings in New South Wales* . . ., George Bennett, arriving at Sydney in December 1841 on his way to New Zealand, set down his feelings about places and people in coastal New South Wales at the time. His disdain of the sandstone heathlands and his liking for an ordered settled landscape reflect the feelings shared by those with a 'colonial vision' of the Australian landscape (Heathcote 1972: 88):

> my first impression was disappointment. I had heard it cited as second only to Rio Janeiro as a Harbour, and I had drawn a fancy picture of lofty mountains and magnificent trees, but I found only gentle hills of sand-stone and dwarf shrubs. The disappointment was my own fault: I had jumbled in my brain a fine harbour with fine scenery. They are distinct, and I have no doubt Sydney still holds its comparative value as a harbour . . . As we entered, it is true on every hill and vale was placed some pretty smiling villa with its garden and green veranda, but the soil around was parched, arid and barren, and nothing but these eternal shrubs, the gum tree and swamp oak stared me in the face — the latter as much resembling our oak as I do an oyster.

He travelled up the coast to Newcastle, and then inland to Maitland:

> The most melancholy thing in nature is a newly settled town such as Maitland, with its would-be streets marked out by palings, and its little allotments for houses similarly subdivided, whilst here and there a wretched wooden tenement occupies about a fourth of an allotment for a house. Not that but there are some good houses scattered about, but the majority of these are Inns and stores. There is no attempt at regularity. It is a mixture of every sort of architecture, but those recognized as such. Then again, the miserable appearance of the half-cleared surrounding country, with its blackened stumps left perhaps for the next generation to uproot.
>
> This indolence as to cultivation may perhaps arise from the badness of the soil, which would ill repay the labor, but why do people make settlements on such land. Sandstone, endless sandstone has haunted me from Cape Howe to Sydney and from Sydney to Maitland. The commons of England, or the bogs of Ireland would afford more produce and more food were the same expense bestowed on them. The trees all look as if they were half blighted, and put one in mind of nothing but the people (currency) of Sydney — a tall cadaverous weedy looking race without stamina or proportion.

Bennett then went south of Sydney where he found the countryside more acceptable:

> The scenery at Wolangong is beautiful, the only part of the Colony I have yet seen to which I can attach that adjective — high mountains and rich vallies covered with lofty tree . . . Still . . . the district of Illawara is overrated, about 5 miles around Wolangong the soil is rich and fertile in the vallies, but after ascending the mountain you come to the same arid parched sandstone country, which only produces a blade of grass beneath the never-varying gum tree.

Macandrew (1877), a young man aboard the clipper *Harbinger,* tried to analyse his feelings about Adelaide:

> Although I was much pleased with Adelaide, there was something wanting — I can't exactly define what — which I

> had been led to expect from the various reports of the old colonists on board. I think that the width of the streets and the comparatively little traffic had something to do with it, after having just left London, with all its busy bustle and noise.

If there is a common theme of dislike about aspects of the colonial scene, it lies perhaps in the untidy landscapes and unfinished and incomplete nature of the urban settlements. In the emigrants' diaries, as in accounts written by others (Heathcote 1976: 42), it is the tidier parklike aspect of the open eucalypt woodlands which appears most acceptable. Hoare (1854) writes that 'the appearance of the land around Port Philip is most beautiful, like a long continuence of Gentlemens' Parks' — a comparison also noted approvingly for the Adelaide Plains by Thomas (1836, in Hope 1968: 114) and the Campbelltown and Illawarra districts of New South Wales by Bennett (1841).

Perhaps the most pathetic case of disappointment at first impressions occurred at Lyttelton in 1850 when the *Charlotte Jane*, one of the first four ships carrying emigrants founding the settlement of Canterbury, sailed into the harbour (Brett 1976(2): 74): 'a number of the women passengers burst into tears when the vessel drew into port. They were crying because they had all dressed themselves in their Sunday best, and there was nowhere to go'.

The Cultural Heritage

Although most settlers in Australia had to endure the long sea voyage from Europe, their collective experience has apparently left little impact on the national character, on the nation's image of itself, or even on the vocabulary of Australian English. Ward (1965) has nothing to say on the experience of the sea voyage as having any influence on the origin of the Australian national mystique. As Bach (1976) mentions: 'Economically and historically indebted to the sea though it be, Australian society gives precious little room to that element in the national mythology it has laboriously contrived to put together over its two centuries of being'. He points out that the influence of the bush has been over-riding, and that large ships have perhaps been seen as symbols of British authority and of a society in which most emigrants found no place. Perhaps the sea voyage from

England, an intensive experience in which most of the emigrants were thrown into close physical contact, was too brief by contrast with the pastoral life, gold-mining or the bush to have any enduring influence on the national character. Can the possible beginnings of a few social attitudes, a few words in use in Australian English such as 'grog', 'killick' and 'muster' (Ramson 1966, 1970), be all that remain in our cultural heritage of the sea voyage to the colonies undertaken by hundreds of thousands of people over the course of half a century or more?

Acknowledgments

I am grateful of Mr Geoff Halsey and Mr R.J. Williams for advice and production of Figure 2.1, and to Tim Wace for help with calculating great circle distances. The Turnbull Library, Wellington; the National Library of Australia, Canberra; and the Capper Press, Christchurch, kindly allowed reproduction of illustrations and quotation from the manuscripts listed in the references. The J.W. Barnicoat MS. is quoted by permission of Mr J.J.G. Barnicoat, Napier, New Zealand, and the C.A. Brown MS. by permission of Keats House, London.

References

Anonymous. 1855. *Blue Jacket* (journal and chronicle of the Blue Waters) MS. National Library, Canberra.

Bach, J.P.S. 1976. *Maritime History of Australia*. Nelson, Melbourne.

Barnicoat, J.D. 1841. Journal aboard the *Lord Auckland* from Gravesend to Nelson 1841-2. MS. Turnbull Library, Wellington.

Bennett, G. 1841. Journal, 1838-45, including a voyage from Plymouth to Sydney aboard ship *Lady Clark* in 1841 (transcribed by B.J. Findlay, 1966). MS. Turnbull Library, Wellington.

Bent, J.H. 1814. Journal of a voyage in the ship *Broxbornebury* from England to Sydney in 1814. MS. National Library, Canberra.

Blainey, G. 1966. *The Tyranny of Distance: How Distance Shaped Australia's History*. Sun Books, Melbourne.

Borrie, W.O. 1955. Australia. Pp.78-117 in O. Handlin and B. Thomas (eds.), *The Positive Contribution by Immigrants*. UNESCO, Paris.

Brett, H. 1976. *White Wings: Fifty Years of Sail in the New Zealand Trade, 1850-1900*. Brett Printing Company, Auckland, 1924; reprinted 1976 Capper Press, Christchurch.

Brown, C.A. 1841. Log book kept on voyage to New Zealand in 1841, aboard *Oriental*. Copy in MS. Turnbull Library, Wellington.

Chapman, G.D. 1972. *Kangaroo Island Shipwrecks*. Roebuck Press, Canberra.

Charlwood, D.E. 1971. *The Wreck of the Loch Ard*. Angus and Robertson, Sydney.

Cunningham, P. 1827. *Two Years in New South Wales* (Second edition). Colburn, London.

Featherston, I.E. 1841. Letter to his brother on arriving at Wellington, 5 June 1841. MS. Turnbull Library, Wellington.

Fell, A. 1926. *A Colonist's Voyage to New Zealand.* Townsend, Exeter; reprinted 1973 Capper Press, Christchurch.

Fischer, G.L. 1962. Captain John Hindmarsh's Letters to George Fife Angas, part I: 1835-1837. *South Australiana* 1: 5-13.

Heathcote, R.L. 1972. The visions of Australia 1770-1970. Pp.77-99 in Rapoport, A. (ed.), *Australia as Human Setting: Approaches to the Designed Environment.* Angus and Robertson, Sydney.

Heathcote, R.L. 1976. Early European perception of the Australian landscape: the first hundred years. Pp. 29-46 in G. Seddon and M. Davis (eds.), *Man and Landscape in Australia.* (Australian UNESCO Committee for Man and the Biosphere, Publication No. 2). Australian Government Publishing Service, Canberra.

Herd, J. 1832. Remarks on Tristan. *Nautical Magazine* (September 1832).

Hoare, F. 1854. Journal aboard *Red Jacket* from Liverpool to Melbourne, May to July 1854. Microfilm: National Library, Canberra.

Hope, P. 1968. *The Voyage of the Africaine.* Heinemann, Melbourne.

Horsburgh, J. 1809. *Directions for Sailing To and From the East Indies, China, New Holland, Cape of Good Hope, and the Interjaceant Ports.* East India Company, London. Also editions 1826 and 1841.

Lubbock, B. 1948. *The Colonial Clippers.* Brown, Son and Ferguson, Glasgow.

Lubbock, B. 1957. Merchantmen under sail 1815-1932. *Mariner's Mirror* 43: 3-18.

Lyle, T.D. 1852. Copy of notes of a voyage to Australia in the ship *Nepaul.* MS. National Library, Canberra.

Macandrew, I.F. 1877. Diary of a voyage aboard the ship *Harbinger* from England to Adelaide in 1876. MS. National Library, Canberra.

Manning, A.W. 1839-40. Journal in the ship *Earl Grey* on a voyage from England to Australia. Photostat: National Library, Canberra.

Martin, A. 1851. Journal of voyage aboard vessel *Cashmere* from London to Auckland. MS. Hocken Library, Dunedin.

Maury, M.F. 1855. *The Physical Geography of the Sea and its Meteorology.* Harper, New York.

Purdy, J. 1844. *Oriental Navigator, Or Sailing Directions for the East Indies, China, Australia, Etc.* Whittle and Holmes, London.

Ramson, W.S. 1966. *Australian English: An Historical Study of the Vocabulary, 1788-1898.* Australian National University Press, Canberra.

Ramson, W.S. (ed.) 1970. *English Transported: Essays on Australasian English.* Australian National University Press, Canberra.

Ricou, J.P. 1872. Diary of the voyage of the Barque *Indus* (Capt. E.P. Ellis) from London to Brisbane. MS. National Library, Canberra.

Taylor, J.M. 1841. Diary aboard *Sir John Falstaff* from Gravesend to Sydney, 1840-1. MS. Turnbull Library, Wellington.

Torlesse, P. 1856. Diary of voyage from Gravesend to Lyttleton aboard vessel *Egmont.* MS. Canterbury Museum, Christchurch.

Ward, R. 1965. *The Australian Legend* (Second edition). Oxford University Press, Melbourne.

Williams, F.L. 1963. *Matthew Fontaine Maury.* Rutgers University Press, New Brunswick.

Chapter 3

The British Expert Cometh

The Imprint of the Individual on Australia's Economic Landscape

*Peter J. Rimmer**

> From innumerable but incomplete facets of an alien world the Expert must try to capture the feeling and the general form of life, and to construct some generality which shall be at once manageable in the world of affairs and yet not too neglectful or distorting of human facts — of the pride of men who own little but their pride in an identity stemming from ancient ways, who would not willingly lose that nor should be robbed of it, but who yet reach out for new values in the strange new world before them. They long to master that perilous navigation from the old to the new, yet fear to lose sight of their ancestral shores. It is for the Expert to help them chart their course . . . (Spate 1965: 13.)

British Made

Since 1788 Australia's economic landscape has, in part, been shaped by a procession of British experts. Although doubts, summarised by Pawson (1977), have been cast on the efficacy of transport's role in economic growth, Alfred Marshall's dictum that 'the dominant economic fact. . . is the development not of manufacturing but of the transport industries' (Robbins 1962: 156) is accepted here for Australia. Its economic landscape reflects the impact of the railway

* *P.J. Rimmer, a graduate of Manchester and Canterbury Universities, is Senior Fellow in the Department of Human Geography, Research School of Pacific Studies, The Australian National University. He was Professor Spate's last appointment to the 'old' Department of Geography.*

and the motor vehicle which Baran and Sweezy (1966: 219) rank, with steam-power, as 'epoch-making innovations' that have

> each produced a radical alteration of economic geography with attendant internal migration and the building of new communities; each required or made possible the production of many new goods and services; each directly or indirectly enlarged the market of a whole range of industrial products.

Although the motor vehicle has propelled the suburbanisation of economic activities in Australia during the twentieth century, attention is concentrated on the railway epoch (1850-1907) because that was the period when British transport experts exerted their greatest influence. Indeed, transport industrialisation during this period was accomplished 'to a large extent with British capital, British materials and equipment, and often by British contractors' (Hobsbawm 1968: 93). However, the focus of attention during this 'British phase' should not be on the railway alone but on its vital connection with the port which facilitated, through improved (and, increasingly, steam-driven) shipping, the critical connection with the British metropole.

Assessments, by geographers, of the railway epoch's impact on Australia's economic landscape have been made largely in descriptive terms best reflected in the proliferation of railway maps and commentaries showing when each link was opened for traffic. In a bid to boost their explanatory content a quantitative approach has been increasingly adopted in which network analysis is used to provide an efficient description of transport systems in terms of their geometric equivalents of points and lines preparatory to 'explaining' them statistically. However, there has been little recognition in this search for spatial regularities in Australia's transport landscape that 'every piece of engineering "hardware" emerges from a dense jungle of human motives and aspirations, some noble and others quite the reverse' (Rolt 1968: v).

Nevertheless, there is the counter-view that the decision-making processes did not create Australia's railway-port systems. It was the need for a more rapid mode of transport that became imperative with the development and expansion of the capitalist system, not merely to satisfy the transport needs of the economy but to relieve British investors of 'capital glut' (Hobsbawm 1968: 90-2). Indeed, Eliot Hurst (1976: 184) observed that

> As technological time collapsed spatial relationships, world markets expanded and capitalism had to break down spatial, national and ethnic barriers. Thus, when developed, the railway was quickly adopted to achieve those ends, as was the steamship . . . it is not surprising that there is such a close correspondence between the rise of capitalism . . . and technological innovations and their adoption, for above all else it is capitalism that shaped Canadian [Australian] and American landscapes.

While there were similarities in the way 'octopus' railways were used to tie hinterlands to commercial and financial interests in the entrepôts (i.e. metropolitan cores) there were differences in emphasis as to how it was achieved.

Interest in North America has centred on how the economic landscape is a product of 'hard-nosed attitudes by entrepreneurs and others in the context of a burgeoning and profitable capitalist economy' (Eliot Hurst 1976: 185). However, government enterprise has been paramount in railway and port development in Australia albeit as a close correlate of private enterprise. The key figure in America is typified either by the 'general entrepreneur', such as Commodore Cornelius Vanderbilt and John M. Forbes, or the 'professional entrepreneur', such as Charles E. Perkins (Cochran 1953), whereas in Australia it was the bureaucrat endeavouring, not always successfully, to ensure that politicians with financial and commercial interests continually kept some social utility in mind (Butlin 1964: 291, 352-7; Rimmer 1975). What Vance (1970: 140) describes as 'exceptional acts' in America involve the individual entrepreneur or firm shaping the 'organization of business that will gather greater individual, as opposed to generic, profits'; in Australia the 'exceptional acts' were often the result of intervention by the British transport expert, notably the engineer, cast in the role of the 'disinterested professional'.

The concentration on the British engineer is not difficult to substantiate. 'A study of the British engineering profession as it emerged in the years following the development of the railways in the 1830s is crucial to the understanding of both economic history and the history of technology' (Bell 1975: vii). Transport was a key employment sector as, according to Coode (1889: 5):

> wherever there is trade — and where is there not in this essentially busy age of ours? — there will of necessity be a demand for the means of transport; and where there is a demand for the means of transport there will equally be a demand for the work of the [British] Civil Engineer.

Much of the demand came from the Colonies 'and many of the most important works [but not all] have either been designed by, or submitted for the opinion of, leading engineers at home' (Coode 1889: 3). These tasks were fulfilled by the 'second generation' of engineers who followed those responsible for major developments in British engineering, particularly in the second quarter of the nineteenth century (Robert Stephenson, Isambard Kingdom Brunel and Joseph Locke), and contributed to their spread throughout the world between 1850 and 1907 when the contractor rather than the 'great engineer' came to dominate (Armytage 1961; Rolt 1970). Thus, an examination of the British transport expert's role during the railway epoch in Australia provides an opportunity to consider his contribution to rendering 'the country accessible in all directions by means of roads, bridges. . . and railways; and [building] lighthouses, breakwaters, docks, and harbours, for the protection and accommodation of . . . home and foreign commerce' (Smiles 1862(1): iii).

In undertaking this task in Australia one immediately faces the same problems that beset Samuel Smiles (1862(1): iii-v) in writing *Lives of the Engineers* (see also Mackay 1905: 247-63; Smiles, A. 1956). There is little published information on the role of British engineers in Australia presumably because, as in Britain, the 'mundane' subject of engineering has been considered too 'vulgar and commonplace' for 'literary men of eminence' to produce biographies. Even if this reaction could be overcome there is still the chore, which Smiles (1862(1): iii-v) hated, of 'burrowing through a vast amount of engineering reports, which, next to law papers, are about the driest possible reading, except to those professionally interested in them'. A history of The Institution of Engineers, Australia (Corbett 1973) has been written but the country has yet to produce its Smiles (mercifully say some). In many cases only unsatisfactory notices, contained in encyclopaedias and biographical dictionaries, have appeared. Thus, 'the vast majority of the men responsible for the engineering achievements of the Industrial Revolution remain unknown figures, their only memorial being very often obituary notices found in the contemporary engineering press' (Bell 1975: vii) — the British *'corps d'élite'*, the 'navvies', not even achieving that solace.

The difficulty of selecting a sample of British engineers who have worked in Australia is obvious given the scant amount of information available. However, Bell (1975) has compiled a *Biographical Index of British Engineers in the 19th Century* based on 3,678 obituary notices in over forty contemporary British engineering journals. An analysis shows that 545 engineers are recorded as having worked outside Britain of whom twenty-five were in Australia — the second largest entry after India with 222 engineers. The list is a misleading guide to the contribution of the British engineers in Australia: it contains 129 entries classified loosely as 'overseas' or 'colonial'; it excludes those whose working life was concentrated in the nineteenth century but perchance died after 1900; it makes no reference to engineers who, though domiciled in Britain, visited various parts of the Empire; and makes no attempt to use non-British journals. Also the index is riddled with people loosely associated with engineering such as 'inventors, some entrepreneurs and company directors, architects, patent agents, etc.' (Bell 1975: viii). Thus, no pretence is made of selecting British engineers in Australia at random to gauge their role in the changes in technology, economic organisation and industrial structure involved in transport industrialisation, or their effect on the people who worked in the industry and the community in general. Instead, the study adopts a more limited objective and selects a series of case studies merely to highlight the varying nature of the British transport expert's role.

The use of individual case studies does raise problems about the choice of approach. Should one adopt Smiles' (1859) method of emphasising the virtues of thrift, industry and free enterprise epitomised in his book *Self-help* which lionised practical men from humble origins who had made good? Or should one follow Hobsbawm's (1968) method in *Industry and Empire* of playing down the role of the expert as a mere agent of international capitalism because the railways and ports were 'natural' developments which advanced technically as demand for their services increased and as more capital was pumped into them? Rather than choose between the Smiles and Hobsbawm strategies three case studies are made of the varying stays of British engineers in Australia to assess which approach is most applicable.

The first study involves a tour of duty by a military engineer, Captain (later Lieutenant-General Sir Andrew) Clarke (1824-1902), that was sufficiently long for him to produce a blueprint for railways in Victoria before moving on to the next level of the Empire's military-bureaucratic hierarchy which had service in India at its pinnacle. A second study illustrates the role of the peripatetic engineer by examining Sir John Coode's (1816-92) two tours of Australia which affected the design of ports and their ultimate fortunes. The third study mirrors the experience of the Empire's technicians who came and spent their life in Australia such as Yorkshireman, John Whitton (1820-98) — the engineer responsible for implementing most of the initial railway development in New South Wales. Collectively, the three studies help destroy Smiles' 'potent myth' of the meteoric rise of men from humble surroundings as Clarke, Coode and Whitton were born into well-to-do families and possessed the scientific education that the heroes of *Self-help* lacked.

Spicy Andrew's Blueprint

Australia's reliance on advisers from the Empire's patrician class is reflected in the role played by Andrew Clarke in the genesis of railway development in Victoria — a task of supplanting the uncertainties and expense of road transport by orderly and effective communications between the coast and the interior gold-fields. Son of an ex-Governor of Western Australia, Clarke had already served in a military-civil capacity in Ireland, Tasmania and New Zealand before becoming Victoria's Surveyor-General in 1853 and Surveyor-General and Commissioner for Lands in the Colony's first Cabinet following the granting of responsible government in 1856. Dubbed by the colonialists 'Spicy Andrew', Clarke had used his 'zeal and energy' to help frame Victoria's constitution, extend municipal government, supervise land sales, initiate scientific and artistic institutions, instigate the continent's first electric telegraph and plan railways in Victoria for three years until his final proposals were submitted in January 1857. Attention is concentrated on analysing his report on railways (Clarke 1856) and its aftermath — the official 'Report' representing the hallmark of the expert (Spate 1965: 10-11).

As a guide to those deciding upon the best means of internal communications the report outlines the reasons for the lack of progress on railway construction, the results of route surveys directed by Clarke for the legislature and the key factors influencing the cost of railways. Rather than repeat a review of the report (Rimmer 1975), interest is centred on three questions relevant to an assessment of Clarke's impact on Australia's economic landscape. Should railways be constructed and operated by public or private enterprise; should the English or American design of railways be followed; which lines should be constructed?

Clarke's response to the first question was unequivocal: railways should be constructed and operated by the government. Victoria did not have the requisite large capitalists living on interest or small investors necessary for successful joint stock enterprise as witnessed by the virtual collapse of companies established to provide trunk lines. Government assistance to these companies, with promotion expenses and grants of land for a central terminus and in 91m strips paralleling the track, was unavailing without adequate capital. Clarke found that private companies could only succeed on a basis which was highly objectionable — a government guarantee of 5 per cent which was 'an extent of encouragement to which the history of associated enterprise in connection with railways affords no parallel' (Clarke 1856: viii). In taking this stance Clarke quoted approvingly from Lardner's (1850) *Railway Economy* which endorsed some government interference with private railway companies because of the monopolistic character of their operations. Thus, Clarke recommended a Belgian-style, semi-autonomous, 'board of control' and unwittingly contributed to a tradition of 'centralised paternalism' in public works that Australia has been unwilling to break, despite the construction of railways being let initially to British contractors and operations being offered unsuccessfully to private enterprise (Serle 1963: 239). While Clarke made much of the lamentable examples of private railways in Britain and North America their designs were different and the Surveyor-General had to recommend which of the two models to follow.

The reply to the second question was equally categorical: government-controlled lines should follow the substantial (and expensive) mode of construction on the 'best English model' rather than the cheapness of the American plan — a decision ushering in the interminable debate as to whether engineering or economic considerations should be paramount. Engineering logic dominated Clarke's recommendation that heavy lines and powerful engines were required, especially as surveys showed the physical difficulties between the coast and the interior were greater than expected. Nevertheless, Sir William Denison, Governor-General of New South Wales and mentor of Clarke, espoused the economy of lightly constructed railways and smaller engines incorporated in the American plan. Clarke was unimpressed and argued that light railways would be 'perfectly useless' in Victoria particularly as the cheapness of American compared with British railways was illusory since the former's advantages stemmed from gratuitous land grants, almost complete exemption from parliamentary and legal expenses, and reduced safety requirements. Whereas American promoters justified their railways in terms of what Denison described as *direct benefits* 'which are measured by the profits of the railway as a speculative investment of capital', railways in Australia would be justified in terms of *indirect benefits* 'which are those resulting to society at large from the use of the railway' (Goodwin 1963: 75). This consideration of the social benefits of railways in terms approximating to the concept of consumer surplus reflects that the railway debate in Australia was at a relatively high analytical pitch considering Dupuit (1844) was an early exponent of cost-benefit analysis. However, Clarke made little use of these analytical insights in selecting individual railway projects in Victoria — the key factors being listed as availability of labour, requisite capital, practicable gradients and sufficiency of traffic.

As the basis for answering the third question on route location Clarke costed seven different permutations of survey links; 'Group 5' was recommended as the most advisable line for the government to undertake first — a central trunk line from Melbourne to Ballarat and a branch to Bendigo (Fig. 3.1). Clarke (1856: xxiv) suggested that this line

> offered all the advantages to the general population which any of the other groups afforded, besides traversing the most fertile and productive districts of the country, a large area of which was still unalineated [*sic*]; and, as a single work, was more likely to be carried out unchecked by a possible failure of labor or means, than when these were divided between two separate lines.

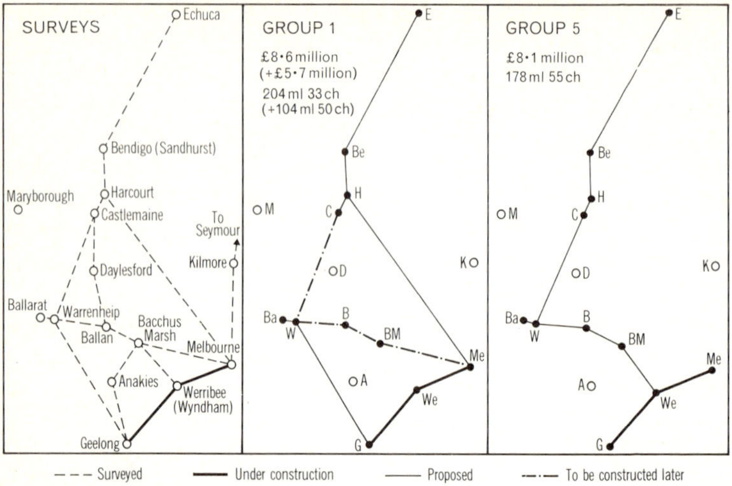

Figure 3.1 Clarke's surveys and costings of Group 1 and Group 5 lines. (*Source:* Rimmer 1975:188.)

The blueprint's requiem

When Clarke presented the railway proposals to the parliament there had been a change of plan as 'Group 1' was substituted involving two lines: Melbourne-Echuca and Geelong-Ballarat (Fig. 3.1). This switch in policy clouded Clarke's contribution as a moulder of Victoria's economic landscape; at least in some contemporary eyes which hitherto had regarded him as 'what the American gentlemen in Martin Chuzzlewit would call "one of the most remarkable men in our country" ' (*Argus* 2 April 1856). The *Argus* (11 August 1858) revised its opinion and held 'Captain CLARKE to a great extent responsible for the cumbrous system of railways which has been adopted'. Clarke's defence was that the land in the interim had been sold in large quantities along the proposed routes (Barkly 1857) — an argument, according to the *Argus* (11 August 1858), that 'resolves itself into little better than a sacrifice of great national interests to the interests of capitalists and landowners'. The two double-tracked lines designed according to the most exacting British standards on the Irish gauge (1.59 m) required a long-term loan of £8 million from the London money market — the most expensive country railways built during the railway epoch (Butlin 1964: 334-51; Blainey 1966: 235). Thus, it appeared that Clarke had succumbed to the pressure 'brought to bear by the inhabitants of particular localities to obtain the adoption of routes for the contemplated lines. . . benefitting alone those parts of the country in which they are themselves interested' (Clarke 1856: xlvi); the very pressure that he had warned the Governor against in his report (see Evans 1857).

It is difficult to assess how much Clarke was stung by the criticism that he had switched from 'disinterested professional' to 'politician'. However, after bringing down the government and feeling unable to be Premier, he confided in a letter to his uncle on 6 July 1859, following a visit to Sir William Denison, 'that on the whole a graceful retreat at this moment is my best policy' and left Australia in August 1858 (Vetch 1905: 65). Although failing to obtain the governorship of Queensland, Clarke enjoyed a distinguished career among the higher echelons of the Imperial Service including being Governor of the Straits Settlements (1873-5) and on the Council of the Viceroy of India (1875-80). Although Clarke had further experience with railways and ports he was not recalled to Australia to advise on technical matters albeit, on occasion, he served as Agent-General for Victoria and Tasmania. Experts, who could purvey distilled wisdom after only a short stay in Australia, were required; one such was Sir John Coode who, despite his formidable reputation did not escape the local predilection for 'expert-knocking' — an aspect of the expert's lot not considered by Spate (1965).

Sir John and the Resident Irish Engineers

When Sir John Coode first visited Australia in 1878 he was already recognised as the leading British, if not the world, authority on harbour works. After an apprenticeship with James Rendel of Plymouth he worked as an engineer on railways in Britain and Spain before succeeding his master as Engineer-in-Chief of the harbour of refuge at Portland, England (1856-73). In addition, Coode had already inspected British ports, harbours and rivers as a member of the Royal Commission on Harbours of Refuge in 1858 and later reported on several of them. He had also examined harbours in the South African colonies and Ceylon; his notable harbour designs included Table Bay and Colombo (Institution of Civil Engineers 1893). After this experience it was not surprising that Coode's services were prized by colonial port authorities as, according to The

Institution of Civil Engineers (1893: 341), by the 1870s he had reduced report compilation and writing to a fine art.

> His power of grasping local conditions was very marked, but he would never commit himself to definite opinion on any scheme until he had before him technical data which could be guaranteed as absolutely correct. For this reason his reports were, as a rule, not made until the surveys and data for which he framed detailed instructions were sent home for consideration.

These meticulous procedures, detailed in the *Age* (7 March 1878), were repeated at Melbourne, Portland, Geelong, Warrnambool, Port Fairy and Lakes Entrance in Victoria and at Murray River, South Australia, during that same year. In 1885 they were repeated at Fremantle and Geraldton (Western Australia); Port Adelaide (South Australia); the Richmond, Clarence, Bellinger and Nambucca Rivers, and Lake Macquarie (New South Wales); and at Mackay and Townsville, and the Fitzroy and Norman Rivers (Queensland). Wollongong, Kiama, Shoalhaven, Darling Harbour and Port Macquarie in New South Wales were also inspected in 1885 but not reported upon by Coode (Institution of Civil Engineers 1893: 339-41).

The need for such a large number of studies was not only to maintain external trade links with the Mother Country but also, in Coode's (1889: 6) opinion, to facilitate internal communications because in

> Our larger Colonies [like Australia], with their great extent of seaboard, as a matter of economy, in fact of necessity, the practice within them has been to extend the roads and railways from the sea-ports on the coast-line back into the interior, with the broad result that the land lines of communication parallel to the coast have been as yet, comparatively speaking, but little developed.

Coode's reports show that he was fully conscious of the need to improve port depths and wharf facilities to accommodate the growing size, number and power of ships even in the most unpromising locations in a bid to expedite trade. A review of the opinions of subsequent observers suggests that Coode's peregrinations had mixed success. For example, Melbourne was regarded as a 'sound concept'

(Kerr 1965); the plans for the Fitzroy River at Rockhampton were endorsed by visiting American expert Lindon Bates (Lewis 1973: 102); Warrnambool was considered to be a 'calamitous failure' (Victoria 1925: 800); Lakes Entrance was declared an 'engineering misconception' (Victoria 1927: 763); and Mackay (Lewis 1973: 70-2) and Fremantle (Bach 1976: 275-6) were not implemented. Other ports, such as Port Fairy (Logan 1967), fell into 'comparative desuetude' not on account of Coode's improvements but because of the extension of the railways and 'the practice of deep freight cutting applied to sea traffic' (Victoria 1925: 828). Rather than discuss each port in detail attention is focused on Melbourne, the sponsor of Coode's 1878 tour and recipient of a second visit in 1885, because it best illustrates the problems of assessing the impact of the short-term expert on Australia's economic landscape.

Port of Melbourne

The inability of the Melbourne Harbor Trust which came into operation in 1877 to resolve the competing claims of the Yarra River and Hobsons Bay as the future focus of the port prompted it to seek outside advice. As there was by then considerable engineering talent domiciled in Victoria 'the propriety of getting an engineer to come to this colony from England' (*Age* 7 March 1878) caused the Trust some anxiety before settling on Coode for a fee of 5,000 guineas (Hoare 1927). The *Argus* helped mollify local opinion by noting (Ruhen 1976: 136) that 'it was right to call in Sir John Coode, but it is due to our resident engineers to point out that he has been called in to confirm their opinions'.

When Coode arrived in Melbourne mail ships anchored at piers near the entrance to the Yarra River which could not accommodate vessels with a draught greater than 3.7 m. In making his inspection of the port and its environs Coode had some remarkable theories propounded to him by 'local experts' who, according to the *Age* (7 March 1878), alluded to his relatively short stay in Melbourne:

> One gentleman came to him and said, 'You cannot know much about the port, as you have only just landed. I have been here for twenty-five years:' He (Sir John) flattered himself that thirty-five years' experience in effecting harbor improvements in nearly all parts of the world was better than twenty-five years' residence in the port of Melbourne . . . However, he

listened to this gentleman of twenty-five years' residence, and the very first statement he made as to a matter of fact he erred in, and he had to tell him that he had got hold of the wrong end of the stick . . .

Having debunked the theories and ascertained the facts as far as possible for himself Coode returned to England with two schemes 'floating in his mind' — direct canal or river improvements and docks.

After receiving additional information Coode (1879) gave unhesitating support in favour of the importance of the existing river course (Fig. 3.2). The report recommended:

(a) a phased suite of three wet docks at the eastern end of the West

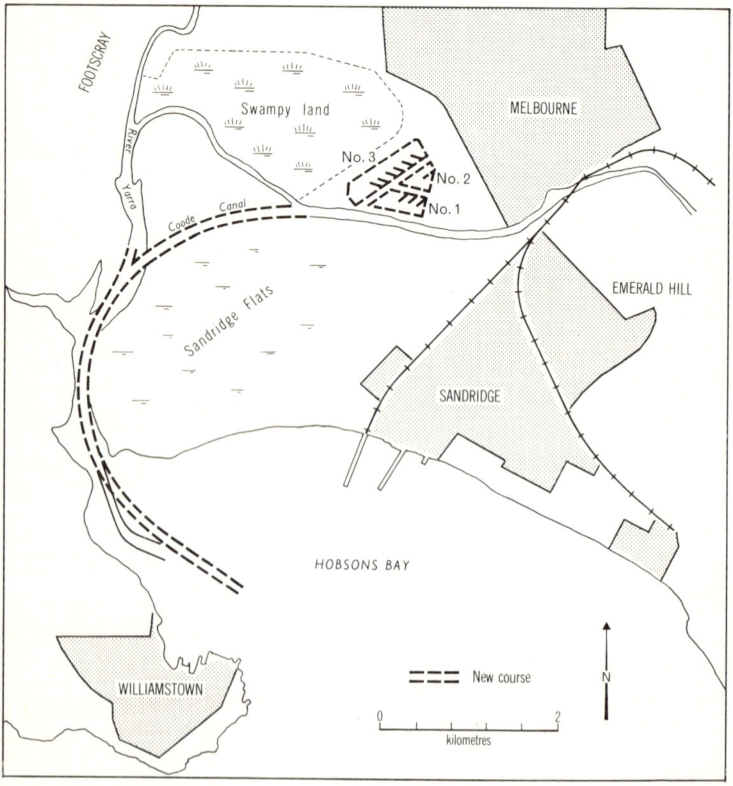

Figure 3.2 Coode's plan for the Port of Melbourne. (*Source:* Melbourne Harbor Trust.)

Melbourne swamp immediately adjacent to the railway terminus (the enclosed docks were to have quays of masonry and concrete supported by subaqueous timber or steel piling on account of the great depth of silt);

(b) the shortening of the river course between these docks and the river mouth by a 'cut' (the Coode Canal) across Fishermens Bend to avoid the 'great detour' (the resulting island became Coode Island).

It was proposed that No. 1 Dock and the river improvements should be undertaken immediately at a combined cost of £1,163,000 to allow the port of Melbourne to berth the largest vessels in the Orient and the Peninsular and Oriental fleets with draughts of between 6.1 m and 7.6 m. The Coode Report was accepted by the Harbor Trust and the river improvements commenced but it was not until 1883 that permission was given to construct No. 1 Dock.

By then the Harbor Commissioners had the opportunity to economise through amendments to Coode's original plan, proposed by the Trust's Irish-born Resident Engineer, Joseph Brady, who had twenty-seven years' experience on railways, ports and waterworks in Australia after working initially on English railways (Kerr 1965). Brady's suggestions were to excavate all three docks as a single dock because No. 1 was considered too small and to substitute the concrete-clad dock walls estimated at £646,400 by red gum timber piles costing £342,000 (Hoare 1927: 75-112). Anxious not to make unscientific amendments to the plan the Commissioners sent Brady's proposals to Britain only for the dock proposal to be accepted and the timber piling to be rejected by Coode's agent. Brady still persisted and his second report, according to Kerr (1965), shook the expert's confidence as Coode (1886) submitted a revised estimate of £1,492,000 for the docks, based on even more elaborate foundations, after paying a second visit to the port in 1885 and receiving further information. As Brady's use of local timber reduced the total cost of Coode's revised scheme (£1,654,000) by £504,000 the Commissioners found their resident engineer's proposal irresistible. Their finding in favour of timber piling was widely regarded as a triumph for 'local' engineers and local materials when Coode's original scheme, as amended by Brady, was completed in 1892. (Kerr 1965, and Bird 1968: 77 differ on the efficiency of Brady's changes.)

However, this 'confrontation' was not the only difference of opinion between Coode and a resident engineer.

Coode's (1878, 1887) designs for an outer harbour at Fremantle, sight unseen in 1877, and revised in 1887 after a visit in 1885, were reported on adversely by the State's Irish-born Chief Engineer C. Y. O'Connor (1891, 1892), who offered an alternative plan. On this occasion the stature of Coode almost prevented what later proved to be a most successful inner harbour scheme from being implemented as the parliamentary debates show the strength of support among politicians for short-term experts (Forrest 1891). Nevertheless O'Connor was of the opinion that if Coode had been in possession of later naval surveys 'he himself would have reached the same conclusion' (Buchanan 1927a,b: 291). What would have happened if Coode had stayed in Australia longer? Spate (1965) did not speculate about the possibility of a 'resident expert'. At what stage is the expert status lost? Would the resident expert have encountered similar experiences to John Whitton during the railway epoch?

Honest John's Endeavour

When John Whitton was officially confirmed as Engineer-in-Chief of the publicly-owned and managed New South Wales Railways on 15 January 1857 there were thirty-seven kilometres of 1.43 m track between Sydney and Liverpool and a further twenty-seven kilometres under construction from Honeysuckle Point (now Civic Station, Newcastle); a not inconsiderable achievement for a capital-scarce pastoral economy comprising small, isolated and scattered settlements. At Whitton's retirement at the end of May 1890 there were 3,457 extra kilometres of track at a cost of £29,800,000 in a coherent system that linked Sydney's merchants with the borders of Victoria and Queensland and, incidentally, squashed any pretensions Newcastle had for a continuing existence independent of the capital's commercial interests. In the process a new economic landscape was created; its artefacts comprising railway stations, bridges, viaducts and lines designed (or approved) by Whitton (Fig. 3.3). The resulting transport network not only channelled flows of coal, wheat, oats, non-ferrous ores, such as tin, and, to a lesser extent, wool for export by firms, settlements and regions within the colony's diversified and fast industrialising economy but helped overcome psychological distance by transforming New South Wales into a modern, centralised state and paving the way for ultimate federation (cf. the description in Mazlish 1965:31, of the railroad's impact on the American imagination). What was Whitton's role in this metamorphosis? Was he a mere functionary of the British railway establishment? Or did he use his mind and ability as an engineer to plant his own distinctive imprint on the economic landscape?

The questions arise from an examination of Whitton's engineering experience in England. At first sight Whitton's credentials seem unremarkable — surveys and preparation of plans for railways, waterworks and a canal (Institution of Civil Engineers 1898). However, Whitton was assistant engineer on the Oxford, Worcester and Wolverhampton Railway — the well-known firm Messrs Peto (Brassey) and Betts was contractor and the chief engineer was the eminent railway authority (later Sir) John Fowler. On 27 March 1856 Whitton was appointed by the President of the Board of Trade in England (Lord Stanley) as Engineer-in-Chief to lay out and superintend the construction of railways in New South

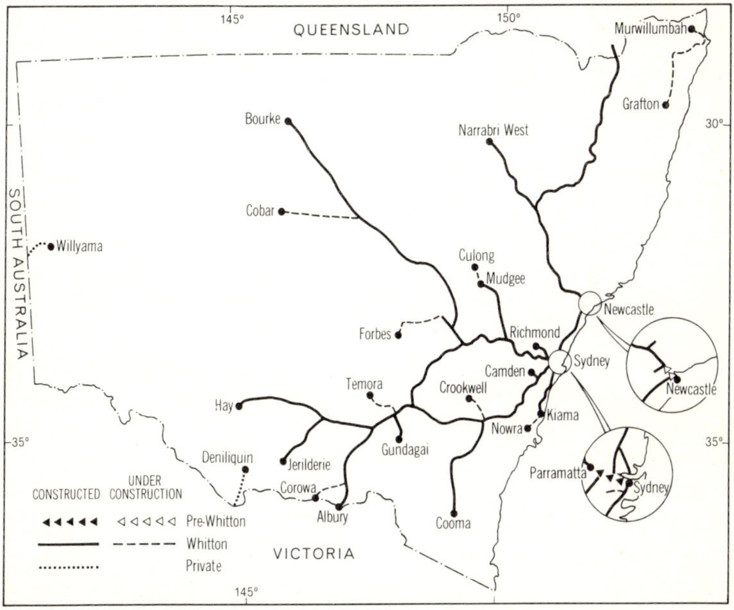

Figure 3.3 Government railways in New South Wales 1890. (*Source:* Public Transport Commission of New South Wales Archives Section.)

Wales under the direction of the government on the recommendation of Sir Moreton Peto. Before leaving for Australia Whitton married Elizabeth Fowler and became Sir John's brother-in-law. These connections with luminaries of the British railway establishment were to dog Whitton in the technical and personal battles that punctuated his career in New South Wales.

Whitton's technical battles are reflected in his reports between 1857 and 1889 which, according to Rae (1898: 9),

> give a minute and lucid description of every mile of the routes followed on each of the lines, their gradients, the different contracts, the number and names of the contractors, and the amount of each contract; in short, the definition of the routes, and the cost of working the railways, during the thirty-three years they were under his charge; and, in addition to these duties, he had charge, for many years, of the permanent way, locomotive and rolling stock branches, and acted as general adviser in most matters connected with the traffic branch.

In particular, the reports show how Whitton, with the aid of E.C. Cracknell, overcame engineering problems partly created by the government's cheese-paring, by using two great zigzags on the Blue Mountains Line which was officially opened on 4 April 1876 — the wonder of the intercolonial picnic even before it was completed (*Sydney Morning Herald* 27 January 1873). Also these reports reflect Whitton's strength of character in:

(a) refusing to sanction 'third class' main trunk lines in Governor Denison's 'economy plan' to construct 6,437 kilometres of cheap, narrow-gauge horse-drawn railways by suspending construction until his term of office had expired and sabotaging the MacLeay (1869, 1870) Select Committee's plans for cheap, narrow-gauge railways by again suspending all surveys until the threat receded with the next turn of the political whirligig;

(b) resisting external pressures from ministers by refusing their appointees to his department (Rae 1898: 12) and popular attempts to mislead exemplified by the rebuttal later of advocacy (*Sydney Morning Herald* 6 June 1873) of cheap lines, narrow gauges and duplicate engines constructed, according to the Engineer-in-Chief, 'on the Siamese principle of "two single gentlemen rolled into one" ';

(c) insisting on high class workmanship and materials instanced by the refusal to replace imported steel rails (£7 per ton) with iron ones manufactured from local ores at Lithgow (£10.5s per ton);
(d) advocating (vainly) the conversion of New South Wales railways to the 1.59 m gauge adopted in Victoria and South Australia, the substitution of Hyde Park for 'draughty' Redfern as the city terminus, and expensive British high speed lines.

Such qualities did not endear themselves to 'log rolling' politicians intent on 'squaring votes' to pursue 'wool lines' when squatting interests prevailed (Laszlo 1956: 37-8). Unwilling to bend with the prevailing political wind, Whitton, in propagating the British railway tradition, had to endure public charges against his probity and innuendoes in parliamentary debates in a concerted bid to prise him from office, his connection with British railway interests providing his opponents with ready-made ammunition.

Little shot was provided by Whitton renewing his connections with Sir Moreton Peto and Co.'s contracting combine in 1858 because the firm left New South Wales in the 1860s after building only a few lines following dissension with the Engineer-in-Chief and the government (Laszlo 1956: 22-3). However, in October 1864 allegations of fraud following 'suspicions in certain quarters' were made against Whitton and his brother-in-law, Sir John Fowler, acting as inspector of railway materials bought in Britain by the New South Wales government; the charges were proved groundless in April 1865 by W. C. Mayne (1866), Agent-General for New South Wales. Although irritating to Fowler these allegations were merely a preliminary skirmish compared with charges against the Engineer-in-Chief during his personal battles with Commissioner Charles A. Goodchap.

The origins of the feud were in a memorandum, allegedly written under the Minister's name by Goodchap, then Secretary, that allowed the Minister to extricate himself from a position in which he had failed to sheet blame for an accident on the Western Line onto Whitton and John Rae, then Commissioner for Railways and Under Secretary for Works (Laszlo 1956: 39-40). As a reward Rae's positions were split and Goodchap became sole commissioner in 1878. Henceforth, the reports of the Engineer-in-Chief and the Commissioner were published separately. However, their reports skirt the feud that influenced the nature of railway development in

New South Wales during the 1880s; inferences have to be drawn from parliamentary debates, public inquiries and the correspondence of the responsible ministers.

On appointment Goodchap immediately sought to shackle Whitton's influence and repay his political benefactors by depriving him of extraneous duties; George R. Cowdery Snr, an ex-colleague on the Oxford, Worcester and Wolverhampton Railway, was made Engineer of Existing Lines with a staff also known to be antipathetic to the Engineer-in-Chief. This ploy had little impact and, when the track doubled from 1,181 kilometres in 1879 to 2,640 kilometres in 1884 in response to the flood of British capital (Butlin 1964: 335), the feud intensified. Such unprecedented growth did not sate 'log rolling' politicians and Whitton was ordered to survey 3,541 kilometres of track in 1883 as ministerial bait for their support in the Assembly. Whitton resisted such schemes by allegedly costing them exorbitantly and refusing to sign plans and books of reference for presentation to the Assembly — a strategy that voided them. As the falsification of the Engineer-in-Chief's estimates was too risky as a regular tactic there were bids to deprive Whitton of his influence and legal responsibility. Various subterfuges were tried including having Whitton elevated to the position of Consulting Engineer with an increase in salary to retain his scarce technical expertise and having the Engineer-in-Chief made subservient to the Commissioner (Laszlo 1956: 53-4). As these manoeuvres failed Whitton was subjected to a New South Wales Royal Commission (1885a, b) into the stability of iron bridges on the existing lines at the instigation of Cowdery; it exonerated Whitton of faulty design and of using inferior materials. Financial mismanagement and undue political influence in railway administration, reported to Parkes by Whitton as early as 5 January 1883 (Parkes 1876-90), ultimately resulted in Goodchap being relieved of his position in 1888, the establishment of a Committee on Public Works to consider future extensions, and the appointment of three railway commissioners — the Chief Commissioner, E. M. G. Eddy, coming from Britain (as did his respective Victorian and Queensland counterparts Richard Speight and John Matheson). By then Whitton had mellowed (or come to terms with local conditions) as he agreed light lines could be built as branches to the main trunks. Whitton retired in 1890 and was succeeded by Henry Deane — an ex-articled pupil of Fowler. But Whitton's trials and tribulations were not over as he had been too successful in pursuing the public interest in New South Wales and unforgiving politicians were poised to strike on his retirement.

A government proposal to retire Whitton on full salary was supported by the Premier, Sir Henry Parkes (1890: 6583-4) who reminded the Assembly that 'John Whitton was an honest servant; and, as he had the power to command immense sums of money, it is easy to see how the country must have benefited by his . . . thoroughly honest service'. Even 'Mr. Goodchap . . . an inconsiderable public servant compared with Mr. Whitton . . . retired with a pension calculated at three-fourths of his salary' (Parkes 1890: 6584). However, the Premier's plea fell largely on deaf ears.

There were those in the Assembly who were genuinely against pensions for public servants. Some members were moved by the fact that they did not 'apply to the man who worked on the railway lines as a fettler' (Haynes 1890: 6588). Whitton's pension also provided an opportunity to embarrass Parkes because as Mr Schey (1890: 6590), an ex-railway porter, noted 'it was embalmed on the pages of *Hansard* that the Premier had a very close, very fraternal, and very peculiar friendship for Mr. Whitton'. However, the main thrust of Schey's attack (1890: 6592) on Whitton was the British connection. It was alleged that Whitton had 'so directed huge orders given from this country to Great Britain that he had succeeded in putting into the pocket of his brother-in-law . . . the sum of £65,000 in twenty-five years'. Schey (1890: 6591) argued that 'credit was due to the then Dibbs Government for disregarding the advice of Mr. Whitton, who wished to have the bridge [over the Hawkesbury] designed and carried out by . . . Sir John Fowler . . . and the result was a saving to this country of £200,000'. At the time of his speech Schey (1890: 6592) also highlighted that 'there was a sum of nearly £4,000 going to Sir John Fowler in connection with the contract just entered into for locomotives' but made no reference to his proposal for a harbour bridge put to Parkes by Whitton in letters of 23 March 1886 and 20 January 1887 (Parkes 1876-90). Nevertheless, Schey's comments, without this extra information, hit home and the Government's proposal was lost and Whitton's pension reduced to £675 — £125 less than the pension enjoyed by the Member for Redfern, Mr Goodchap. Sick and humiliated Whitton died in 1898 feeling 'that his services . . . were unknown to the present generation, and unappreciated by the public' (Rae 1898: 1). Was this the way that Australian capitalism rewarded thrift and industry once the services of the expert were no longer required?

Guest of Honour

These three case studies highlight the difficulty of choosing between the Smiles (1859) and Hobsbawm (1968) approaches; the former would *overrate* the expert's contribution in shaping Australia's economic landscape by overstressing personal attributes and the latter would *underrate* it through simply making the development of railways and ports one of common sense technical adaptation by men out for a profit — it was not as smooth and 'natural' as Hobsbawm suggests. A compromise is possible by accepting Harvie's argument that the *opportunity* for the expert to affect Australia's landscape depended on external economic forces but that the *nature* of his impact was an expression of the engineer's mind and ability (Open University 1970: 47-9).

Since the railway epoch in Australia the procession of British and other overseas experts has continued unabated. Without strict import controls the expert has become 'one of the figures of fun of our day, and often doubtless deservedly: there are simple problems and simple-minded Experts content (not always wrongly) with the good intentions and facile recommendations of our improving age' (Spate 1965: 13). Whether the experts are transnational employees on short tours, itinerant academics on the university lecturing circuit, or long-term recruits to the public service, Andrew Clarke, Sir John Coode and John Whitton should remind them that they have in their hands, according to Spate (1965: 13-14) whose career embodied all three types of expert, 'not just the "progress" but the real happiness of thousands or tens of thousands; and that is not so funny'.

Acknowledgments

Grateful acknowledgment is made of assistance from Barbara Banks (The Australian National University); Gwen Carroll (Menzies Library); John Forsyth (Public Transport Commission of New South Wales Archives Section); Melbourne Harbor Trust and other harbour authorities; Mitchell Library; Latrobe Library; Victorian Public Record Office; and The National Library of Australia.

References

Armytage, W. H. G. 1961. *A Social History of Engineering*. Faber, London.

Bach, J. 1976. *A Maritime History of Australia*. Nelson, Melbourne.

Baran, P. A. and Sweezy, P.M. 1966. *Monopoly Capital*. Monthly Review Press, New York.

Barkly, H. 1857. Governor of Victoria to Secretary of State Despatch No. 117, 2 December 1857. Mitchell Library, Sydney.

Bell, S.P. 1975. *A Biographical Index of British Engineers in the 19th Century*. Garland, New York.

Bird, J. 1968. *Seaport Gateways of Australia*. Oxford University Press, London.

Blainey, G. 1966. *The Tyranny of Distance: How Distance Shaped Australia's History*. Sun Books, Melbourne.

Buchanan, G. 1927a,b. Report on transport in Australia with special reference to port and harbor facilities. *Commonwealth of Australia Papers Presented to Parliament*, 1926-8(5), No. 86: 81-239; No. 108: 241-99.

Butlin, N.G. 1964. *Investment in Australian Economic Development 1861-1900*. Cambridge University Press, Cambridge.

Clarke, A. 1856. Report of Andrew Clarke R.E., M.P., Surveyor-General upon railways; with appendices. *Victoria Papers Presented to Both Houses of Parliament*, 1856-57(4), No. 33.

Cochran, T.C. 1953. *Railroad Leaders 1845-1890: The Business Mind in Action*. Harvard University Press, Cambridge, Mass.

Coode, J. 1878. Report upon the question of harbor improvements at Fremantle. *Western Australia Votes and Proceedings of the Legislative Council . . . with Papers Presented*, 1878, No. 26.

Coode, J. 1879. *Report on Works of Improvement*. Melbourne Harbor Trust. Melbourne.

Coode, J. 1886. *Report 31 May 1886* [to Melbourne Harbor Trust Commissioners]. Melbourne Harbor Trust, Melbourne.

Coode, J. 1887. Report on the question of harbor works at Fremantle. *Western Australia Votes and Proceedings of the Legislative Council . . . with Papers Presented*, 1887, No. 18.

Coode, J. 1889. Address of Sir John Coode, President [British Colonies as fields for the employment of the civil engineer — past — present — and future]. *Minutes of Proceedings The Institution of Civil Engineers* 96: 1-40

Corbett, A.H. 1973. *The Institution of Engineers Australia: A History of the First Fifty Years 1919-1969*. Institution of Engineers, Australia with Angus and Robertson, Sydney.

Dupuit, J. 1844. De la mesure de l'utilité des travaux publics. *Annales des Ponts et Chaussées* 8 (Second Series), trans. R. H. Barback as 'On the measurement of the utility of public works' for *International Economic Papers*, 1952(2): 83-110. Reprinted 1968 as 'Public works and the consumer', pp. 19-57 in D. Munby (ed.), *Transport: Selected Readings*. Penguin, Harmondsworth.

Eliot Hurst, M.E. 1976. The railway epoch and the North American landscape. Pp. 183-206 in J.W. Watson and T. O'Riordan (eds.), *The American Environment: Perceptions and Policies*. Wiley, London.

Evans, G.S. 1857. Railways loan Bill. Pp. 1271-2 in *The Victorian Hansard*, 1856-57. 1856-57.

Forrest, J. 1891. Harbor works at Fremantle. Pp. 236-7 in *Western Australia Parliamentary Debates*, 1890-1.

Goodwin, C. D. 1963. Economists and railways in colonial Australia. *Journal of Transport History* 6 (Old Series): 65-86.

Haynes, J. 1890. Supply: pension to Mr. Whitton. Pp. 6586-7 in *New South Wales Parliamentary Debates,* 1890.

Hoare, B. 1927. *Jubilee History of the Melbourne Harbor Trust.* Melbourne Harbor Trust, Melbourne.

Hobsbawm, E. J. 1968. *Industry and Empire: An Economic History of Britain Since 1750.* Weidenfeld and Nicolson, London.

Institution of Civil Engineers, 1893. Obituary notice: Sir John Coode, KCMG. *Minutes of Proceedings Institution of Civil Engineers* 113: 334-43.

Institution of Civil Engineers, 1898. Obituary notice: John Whitton. *Minutes of Proceedings Institution of Civil Engineers* 132: 393-4.

Kerr, C. F. 1965. The man from County Fermanagh. *Port of Melbourne Quarterly* July-September: 33-7.

Lardner, D. 1850. *Railway Economy: A Treatise on the New Art of Transport. . .* Taylor, Walton and Maberly, London.

Laszlo, I. M. 1956. Railway policies and development in northern New South Wales 1846-1889. Unpublished M. A. thesis, University of New England, Armidale.

Lewis, G. 1973. *A History of the Ports of Queensland: A Study in Economic Nationalism.* University of Queensland Press, St Lucia, Queensland.

Logan, W.S. 1967. Victoria's southwestern ports. Unpublished M.A. thesis, University of Melbourne, Melbourne.

Mackay, T. (ed.) 1905. *The Autobiography of Samuel Smiles, LL.D.* Murray, London.

MacLeay, W.J. 1869. Progress report from the select committee on railway extension; together with the proceedings . . . minutes of evidence, and appendix. *New South Wales Votes and Proceedings of the Legislative Assembly,* 1869 (2): 65-92.

MacLeay, W.J. 1870. Report from the select committee on railway extension; together with the proceedings . . . minutes of evidence, and appendix. *New South Wales Votes and Proceedings of the Legislative Assembly,* 1870 (2): 147-238.

Mayne, W.C. 1866. Report, in Railways: Report of Commissioner for Railways — Appendix to *New South Wales Parliamentary Papers* No. 11A: 673-4. Government Printer, Sydney.

Mazlish, B. (ed.) 1965. *The Railroad and the Space Program: An Exploration in Historical Analogy.* MIT Press, Cambridge, Mass.

New South Wales 1885a. Railway bridges inquiry commission (final report). *New South Wales Votes and Proceedings of the Legislative Assembly,* 1885(2): 235-55.

New South Wales 1885b. Report of the Royal Commission appointed . . . to inquire into the stability of certain iron bridges constructed on the existing lines of railway . . . together with minutes of evidence, diagrams and appendices. *New South Wales Votes and Proceedings of the Legislative Assembly,* 1885-6(7): 265-484.

O'Connor, C.Y. 1891. Fremantle harbor works: report by the Engineer-in-Chief upon the entrance to the estuary at the mouth of the River Swan or at Rocky Bay. *Western Australia Votes and Proceedings of the Parliament,* 1891-2, No. A2.

O'Connor, C.Y. 1892. Evidence. Pp. 9-21 in Report of the joint select committee of the Legislative Council and Legislative Assembly on the question of harbor works at Fremantle together with the proceedings of the committee, minutes of evidence, and appendix. *Western Australia Votes and Proceedings of the Parliament,* 1891-2, No. A9.

Open University, 1970. *The Industrialisation Process 1830-1914*. (Humanities: A Foundation Course Units 29 and 30) The Open University Press, Bletchley.

Parkes, H. 1876-90. Parkes correspondence. Mitchell Library (A 912, A 913, 42: 336-56; 43: 493-507, 512-13).

Parkes, H. 1890. Supply: pension to Mr Whitton. Pp. 6582-5 in *New South Wales Parliamentary Debates*, 1890.

Pawson, E. 1977. *Transport and Economy: The Turnpike Roads of Eighteenth Century Britain*. Academic Press, London.

Rae, J. 1898. *Thirty-five Years on the New South Wales Railways: The Work of the Late Mr John Whitton, C.E. Engineer-in-Chief of the Railways of New South Wales*. Builder Printing Works, Sydney.

Rimmer, P.J. 1975. Politicians, public servants and petitioners: aspects of transport in Australia 1851-1901. Pp. 182-225 in J.M. Powell and M. Williams (eds.), *Australian Space Australian Time: Geographical Perspectives*. Oxford University Press, Melbourne.

Robbins, M. 1962. *The Railway Age*. Routledge and Kegan Paul, London.

Rolt, L.T.C. 1968. Foreword. Pp. v-viii in S. Smiles, *Lives of the Engineers . . . Volume 1*. David and Charles Reprints, Newton Abbot.

Rolt, L.T.C. 1970. *Victorian Engineering*. Allen Lane, The Penguin Press, London.

Ruhen, O. 1976. *Port of Melbourne 1835-1976*. Cassell, Melbourne.

Schey, W.F. 1890. Supply: pension to Mr Whitton. Pp. 6590-2 in *New South Wales Parliamentary Papers, 1890*.

Serle, G. 1963. *The Golden Age: A History of the Colony of Victoria, 1851-1861*. Melbourne University Press, Melbourne.

Smiles, A. 1956. *Samuel Smiles and his Surroundings*. Hale, London.

Smiles, S. 1859. *Self-help; with Illustrations of Conduct and Perseverance . . .* Murray, London.

Smiles, S. 1862. *Lives of the Engineers . . .* (3 vols). Murray, London.

Spate, O.H.K. 1965. On being an expert. Pp. 3-14 in O.H.K. Spate, *Let Me Enjoy: Essays, Partly Geographical*. Australian National University Press, Canberra.

Vance, J.E. 1970. *The Merchant's World: The Geography of Wholesaling*. Prentice-Hall, Englewood Cliffs, N.J.

Vetch, R.H. (ed.) 1905. *Life of Lieut.-General the Hon Sir Andrew Clarke G.C.M.G. C.B. C.I.E. Colonial Commandant of Royal Engineers Agent-General of Victoria, Australia*. Murray, London.

Victoria, 1925. Royal Commission on Victorian outer ports, 1925, First-Fourth Progress Reports. *Victoria Papers Presented to Parliament* 1925(2): 723-836.

Victoria, 1927. Royal Commission on Victorian outer ports, 1927, Fifth-Seventh Progress Reports. *Victoria Papers Presented to Parliament* 1927(2): 749-826.

Chapter 4

'Spoil of the Old World's Farthest End'

Does Distance Tyrannise Science?

*J. N. Jennings**

Even when summoned half-way round the world to advise government in Australia, the technical expert has often been subject to misunderstanding and unfavourable publicity and his advice has sometimes become a political football as is illustrated by Rimmer's essay in this book. The visiting scholar may well escape these vicissitudes but other dangers beset him.

If he is distinguished enough, certain of his pronouncements, sometimes based on brief observations, may attract attention beyond by far the time when they warrant it or even because they exemplify the pitfalls which even great minds cannot always escape. This has been the plight, for example, of Charles Darwin in his mistaken assessment of the bottleneck gorges of the Blue Mountains. 'To attribute these hollows to alluvial action would be preposterous' he wrote in 1845, and went on to attribute them to an obscure process of irregular accumulation on the sea bottom. This explanation is so ill-founded that one wonders whether Charles Lyell himself, whose influence lay behind Darwin's implicit rejection of Huttonian theory, was satisfied with it. But Darwin's erroneous view on this matter has continued to be quoted to this day, perhaps in part because the

* *J.N. Jennings graduated from The University of Cambridge where he was supervised briefly by O.H.K. Spate. He joined the Department of Geography of The Australian National University in 1952 under Professor Spate as Reader (later Professorial Fellow) in Geomorphology. In retirement he is now Visiting Fellow in the Department of Biogeography and Geomorphology.*

weighty rhythms of his long Victorian sentences have an unaccustomed dignity at a time when many scientific authors are scarcely capable of more than the banal 'subject-verb-object' and, even if they are, most scientific editors do not allow them to venture beyond it.

Such a fate was unlikely to have befallen the Czech geographer, Daneš, for he was not as eminent as Darwin — there are few such — and also he came from a land little known and not important to most Australians. Nevertheless he made many contributions to Australian geography. Here an attempt will be made to evaluate them and, particularly, to discern any effects on his writings and on their reception here and abroad of a relatively short sojourn in a country separated from his own not only by great distance and natural contrasts but also by cultural barriers of comparable power.

Daneš' Life

Though a citizen of Prague, Jiří (George) V. Daneš (1880-1928) was country-born in Bohemia and through his love of nature gravitated to geography even at gymnasium (Dvorský 1928). From the Charles Ferdinand University at Prague, where his studies ranged from geology through geography to history and his dissertation was devoted to population distribution in Herzegovina, he moved to the Geographical Institute at the University of Berlin under Richthofen. Already he had become acquainted with Professor J. Cvijić, the great Croatian geographer, and this no doubt encouraged him into the study of karst which became the most important single thread in his scholarship. His habilitation thesis was on the geomorphology of the lower valley of the River Neretva and what later constituted Yugoslavia became his second homeland. In 1906 he was appointed lecturer at Prague University yet he had already travelled widely in North America on two long visits, making a special study of Jamaica. Then in 1909 he left with a botanist colleague for fifteen months travel in Java and Australia.

In Australia, his expeditions were extensive in northern Queensland — coast, Eastern Uplands, Western Plains and the Barkly Tableland — where he concerned himself with the Aborigines as well as with geomorphology and European settlements. Afterwards he made brief visits to many other places, especially limestone areas near Rockhampton, to the west and southwest of Sydney including Jenolan, and in the southwest of Western Australia.

The years 1911-16 saw him rising academically at his Alma Mater, writing much and travelling widely in Europe. Then military service took him to Bosnia where, on the collapse of the Austro-Hungarian Empire, he organised a Czech Legion. After the War his return to his university was short-lived because late in 1919 he was appointed consul-general in Sydney by the independent government of the new state of Czechoslovakia. During his two years in Sydney, he made many more journeys in Australia where he was now more concerned with human affairs than with the natural scene, though he did add to his tally of karst areas by crossing the Nullarbor Plain by train and visiting Buchan in Victoria. During his tour of duty he also made a trip to Papua and Australian New Guinea.

From 1923 to 1927, his academic seniority kept him busy in Europe and then he crossed the Atlantic for the third time to the U.S.A. where he lectured in many places and studied limestone regions galore. There, on his way to the new phenomenon of a film town, Hollywood, the wanderings of this inveterate scientific traveller were cut short by a fatal motor accident.

It is not the present purpose to assess Daneš' lifework as a scholar in its entirety, still less his other activities. Suffice it to say that he was regarded as an outstanding man in his own country and he had a high reputation in Europe and North America, which owed not a little to his command of languages. However, his international academic repute rested primarily on his contributions to the study of karst in which field no one in his time had a more broadly based experience (Brandt 1928).

Australian Karst

It is appropriate therefore to begin with Daneš' work on karst in Australia and one theme of this analysis immediately surfaces — neglect of his writings in this country. In this book it is nice to follow O. H. K. Spate in quoting McAuley (1946) once more:

> William Jantszoon may have been the first
> To set a bootprint on our shores, but he
> Disqualified himself by being Dutch
> From celebration in our history . . .

Yet it would be misleading if nationalism were taken to be the root in Daneš' case; science is less afflicted by this bias than most human activities. Sheer ignorance through inaccessibility has been the cause since of the three papers he wrote on Australian karst only one was published in this country (1910c). The second (1916) appeared in German in the main Bohemian scientific journal, whilst the third (1924a), though in English, was included in the *Festschrift* to Jovan Cvijić. The Yugoslavs did their best, offering these essays in the main languages of science, English, French and German, but Belgrade was not a place of publication conducive to world-wide distribution. The situation is worse than this indicates because the paper published in English in Australia was written and delivered to the Royal Society of Queensland in Brisbane immediately after his tiring journeyings through the outback. Not surprisingly this account is less full and valuable than the later papers composed when his ideas had had time to mature. In consequence his contribution in this field has been less appreciated in this country than it deserves (e.g. David and Browne 1950) and than it has been abroad, even in countries where the language barriers are no less (cf. Maksimovich 1962). This oversight has reached amongst the ranks of Australian speleologists with a special interest in this aspect of geomorphology (e.g. Hamilton-Smith 1966) so that it received mild reproof (Jennings 1966).

The fact that the Russian, Maksimovich, relied in 1963 for his own synthesis of Australian karst on Daneš' accounts of Chillagoe and the Barkly Tableland and that Australian speleologists have only recently advanced beyond them reflect the fact that little work has been done subsequent to the Czech's, in itself a consequence of few people living in a large and harsh land. But they are also a measure of Daneš' physical capacity for exploration — he descended 73 m in Nowranie Cave near Camooweal in 1910 — and his *informed* observational powers. At that time there was most ferment of ideas about karst in continental Europe and Daneš was in the thick of it. These are considerations which explain the comment in 1966 that 'Daneš' work on Chillagoe will usefully promote discussion today' (Jennings 1966).

Williams (1978) is much concerned with the contrast between pedimented tower karst like that at Chillagoe and in the Limestone Ranges of West Kimberley on the one hand and polygonal conekarst of Papua New Guinea and Waitomo in New Zealand on the other. This contrast is the one that Daneš himself recognised in descriptive terms between Chillagoe and the Goeneng Sewoe karst in Java.

Though Daneš did not identify pediments as such at Chillagoe, he did argue that the valleys there had reached a broad form sooner in that climate than they did in a 'normal' humid one, showing he had grasped the essential point. Moreover, he stressed the absence of dolines other than collapse dolines within the towers there. Likewise he enlarged on the presence of much tufa building along the streams at Chillagoe and in the Barkly Tableland, especially on its northern fall. This corresponds with what has been found in West Kimberley (Jennings and Sweeting 1963), and Williams (1978) has valuably enlarged our appreciation of flood-plain deposition of carbonate in the North. This modern awareness has had the stimulus of the pioneer's writings.

Daneš (1916) argued that the caves in the Chillagoe towers were old, mainly because of daylight holes (gaps in the roof) and collapse dolines. This has been sustained by the recognition of phreatic (water-filled) solution features to a considerable height in the towers. Thus the caves formed either before the towers were etched out or simultaneously with that development (Wilson 1975; Ford 1978). Even more recently, Robinson (1978) has pointed to geological evidence suggesting that the towers already existed in Tertiary times at least but probably in the Triassic. Also Daneš thought that fallen block piles at entrances had shut out streams from the surrounding plains which had earlier created the caves but this idea must be qualified substantially because phreatic origins for the caves imply that the lateral entry of surface streams into them has been a latter-day and minor affair.

Daneš went most astray in his interpretation of the Lion's Head (now Dome Rock) and its neighbourhood just north of Chillagoe, where smooth surfaces contrast with the intricate minor surface solution sculpture typical of most of this karst. He attributed the smoothing to former river abrasion, including the formation of rock mills. The latter are in fact solution wells in which gravels from terrain on other rocks have survived deceptively (Jennings 1966) and the general contrast in weathering is now interpreted in terms of different limestone lithologies (Wilson 1975; Marker 1977).

Another of the Czech visitor's arguments was that speleothems (such as stalactites and stalagmites) were no longer forming in Chillagoe caves and belonged to a former wetter climate. With strong alternation of wet and dry seasons at present, it is not easy on present evidence to be sure of this inference. Palaeoclimatic evidence published so far (Bowler *et al.* 1976) points to swings towards drier climates in the Pleistocene. However, if the caves go back into the Tertiary, the likelihood of former wetter phases is enhanced. The point worthy of note here is that this issue, raised by Daneš sixty years ago, is still extant.

In his broad review of karst in Australia, Daneš (1924a) dealt mainly in terms of structure and stage and divided the areas into three groups.

In Tasmania, the Lofty-Flinders Ranges of South Australia and the Eastern Uplands, there are only small pockets of tectonically disturbed and often metamorphosed carbonate rocks of Lower Palaeozoic and Proterozoic age. These are dominated by runoff from surrounding impervious rocks, which imposes normal valleys across them and tends readily to result in multi-level cave systems that range downwards from old, abandoned progenitors to the active generation of river passages. Both of these traits have been underlined afresh in the light of modern knowledge (Jennings 1977; Williams 1978).

Extensive tablelands and plains of quasi-horizontally bedded carbonate rocks which form Daneš' second group include the Cambrian Barkly Tableland, the Eocene-Miocene Nullarbor Plain and the Oligocene Southeast Province of South Australia. His remarks relate chiefly to the first of these of which his personal knowledge was greatest. Here it is relevant to note that he recognised the importance of duricrust. In his view laterite and silcrete were largely responsible for the modest development of karst features: the youthful karst stage achieved was due to the inhibiting effect of duricrust on hydrological evolution, a concept still insufficiently explored today.

Of the third group, the Pleistocene dune limestones, Daneš inspected the belt from Perth to Cape Leeuwin, the best developed in terms of karst, but did not seize on the intriguing points that here karst processes are responsible for forming both the rock and the landforms and that to a degree these must have happened simultaneously (Jennings 1968).

Though he did refer to climatic factors, Daneš gave no great thought to their role. In this he was typical of his time. Contemporaries, such as Grund, saw no reason why karst landform assemblages from all over the world should not be employed to construct a single cyclical scheme of karst evolution on the model of W. M. Davis' geographical cycles of normal, desert and glacial climate. Daneš' reports from Jamaica, Java and northern Australia were especially quarried to this end. This approach came to be much criticised later when the notion of climatic geomorphology flourished; Lehmann (1954) was one of the leading protagonists of this standpoint, having followed in Daneš' footsteps in both Java and Jamaica. He criticised earlier karst studies, such as Grund's, because of their implicit assumptions that processes operated similarly whatever the climate and that the course of karst landform change was the same everywhere to the extent that it depended on climate.

Further, Lehmann (1960) argued that terminology was partly responsible for this fallacy. Important karst landform terms were conceived in relation to the circumstances of the Dinaric karst in Yugoslavia and then applied by observers such as Daneš to features in distant karsts which were significantly different. The prime example Lehmann quoted is the difference between the funnel or bowl-shaped doline of the karst proper and the star-shaped tropical cockpit, which Daneš failed to distinguish, as he might have done, by using the local Jamaican name, 'cockpit'. When Daneš used the term 'doline' for Javan or Jamaican features, he was alert to some of their differences from dolines in Europe, but his readers, such as Grund, blinded by their preconceptions as to what the terms meant, built tropical 'dolines' into false evolutionary schemes — the inverse of *nihil ex Africa semper aliquid novi*.

The wheel has turned again and today it is climatic geomorphology which is under scrutiny. The nature and arrangement of the rocks have undoubtedly been undervalued in karst interpretation in recent decades (Sweeting 1973). If Daneš had known the Limestone Ranges of West Kimberley, he is likely to have placed them in his second karst group because of their similarity in some respects to the Barkly Tableland. Recently Brook and Ford (1976) have attacked the climatic control school on the basis of a strong analogy between the North Nahanni tundra/boreal karst of northwest Canada and the Limestone Ranges whilst Williams (1978) attributed the differences between the temperate Nullarbor Plain and the tropical Limestone Ranges not to climate but to lithological differences. Patently we cannot yet dispense with Daneš' karst writings.

Lakes and Rivers in Northeastern Queensland

Only recently did Daneš' Australian karst studies find an audience here of scientists engaged in this aspect of geomorphology. His more numerous papers dealing with other aspects of northeastern Queensland, notably the relations of rivers and lakes to the geomorphic evolution, were contributions to themes to which his Australian contemporaries were at the time committed. Besides more of his writings in this direction were published in English (1910a, 1910b, 1911), the others in French (1912a, 1912b) — perhaps slightly more accessible to the Australian scientific community. There are two other papers (1912c; 1913) of a semi-popular, descriptive nature which draw on the materials used more penetratingly in the previous articles.

The essence of Daneš work in this field is as follows. Shallow, closed drainage basins set in duricrusted terrain like those of Lakes Buchanan and Galilee survive from a Tertiary surface of low relief close to sea level with dominantly meridional drainage along the structural grain but including many such closed basins orientated sympathetically. The surviving lakes characteristically line the Great Divide, hence his comment that 'The people speak and believe about the Dividing Range, although it would be necessary to take refuge in a spirit level for observation of some rise in the country'. This is reminiscent of Griffith Taylor's remark that a part of the so-called Great Dividing Range in Queensland is named Duck Flat.

Then in the late Tertiary, faulting disrupted this surface, lowering the continental shelf on which the Great Barrier Reef has since grown and elevating what are now the Eastern Uplands. The coastal ranges and intervening coastal corridors were also a product of this phase of epeirogeny and faulting. In consequence rivers flowing eastward and westward eroded headwards into the uplands, captured parts of the former meridional rivers and breached most of the internal drainage basins. Waterfalls, boathook bends on the rivers and former river courses marked by alluvial gravels are part of the evidence for such captures as those of the upper Mitchell River by the Barron River and of Prairie Creek and other streams by the River Flinders.

Closely related discussions were in fashion at this time. Indeed a skeletal suggestion had already been put forward by Hedley and Taylor (1907) that the Barron had captured the upper Mitchell and this was followed by Poole (1910) with a documented case. Contributions which overlapped in content with Daneš' such as those of Andrews (1910), Taylor (1911), and Hedley (1911) appeared after his two 1910 papers on this subject but before those of 1911, 1912a and 1912b. There was personal communication between Daneš and Andrews, Taylor and Hedley (see endnote to Daneš 1912b) and Daneš' 1911 article is essentially a critique of the similarities and differences between his own views and those of the other three. There is no need to pursue this in substance here but it is apposite to note that such interchange between a visiting and a home scientist is what should happen but does not always take place. In this field Daneš seems to have avoided completely the trap of failing to familiarise himself with what had already emerged in this distant land on the themes he was pursuing. Contemporaneously on his home ground, Taylor did not do as well because Hedley (1911) acidly comments with regard to an associated matter, 'Apparently unacquainted with the studies of his predecessors, Mr. T. G. Taylor, half a century afterwards, came to the same conclusion'.

In the 1920s and 1930s there continued to be writing about the hydrography, old surfaces and related tectonics of the Eastern Uplands in Queensland, chiefly because of interest in the foundations of the Great Barrier Reef. Consequently there is more reference to Daneš' writings about these matters than to his karst papers. He is mentioned with approval by both protagonists (Jardine 1925a,b; Stanley 1927; Süssmilch 1938) and antagonists (Marks 1913, 1924, 1931; Browne 1934) of the dominant views of Andrews and Taylor. Inveighing against chair and map-bound physiographers, Marks (1924) praises Daneš for his 'eight months field work, much of it far from beaten tracks' and for the way he 'clearly distinguishes between observations of fact and theoretical deductions therefrom'. Likewise whilst disagreeing with him on some things, Jardine (1925b) speaks more approvingly of Daneš as a predecessor than of any other. The most interesting touch of all (in the light of his earlier stricture quoted above) is Hedley's postscript to his own 1925 paper:

> Since the preceding article went to press, I have been reminded of an important memoir by Dr J. V. Daneš . . . The subject of

> my essay is here treated with greater skill, knowledge and thoroughness than I could command . . . At this point I considered the withdrawal of my article but refrained from doing so on the ground that it may convey information to Australian readers to whom the memoir of my friend is not accessible.

The article in question is in French in a French journal, so that Hedley's final decision was not without grounds in the light of common Australian scientific practice.

There followed a long period when this field of study became inactive but there has been a change of late for two reasons above all. Improved methods of dating rocks are demonstrating that many tectonic events and landforms are much older than was previously conceived. In addition, continental drift and plate tectonics mean that the entire geography of the continent has changed since some of the earlier forms were fashioned. This has revived controversy about the nature of upland surfaces and the explanation of river patterns (e.g. Young 1977). With this newly kindled interest there is certainly need to evaluate afresh early contributions to this field to clarify the issues. This applies, for example, to Daneš' claim that closed basins were characteristic of the surface of eastern Australia prior to Tertiary uplift. It is true that Jardine (1925b) stated that the evidence for such a former basin was lacking in the Russell, Johnstone and Barron catchments and, whilst approving Daneš' rejection of warping as a cause of the closed basins which do exist, Browne (1934) cast implicit doubt on the origin that Daneš envisaged for them himself. So this question needs taking up once more in the light, for example, of the great age shown for the similar Lake George basin in New South Wales in recent unpublished palaeomagnetic and stratigraphic work by Opdyke, Bowler and Singh. As Ollier (1978) has claimed, it is important to pursue such questions not only for their intrinsic value but also because of the help they may give to the decipherment of tectonic history.

The Peopling of Australia

As one of his obituarists (Dovorský 1928) stressed, Daneš was an all round geographer, his last as well as his first contribution dealing with human geography. All this side of his work, including that part which relates to Australia, is in his native tongue. Fortunately two of the more important papers were also published in English translation and the gist of the others was communicated to me by Henry Polach.

Daneš' book (1924b) on the origin and extinction of the Aborigines is remarkable for its balance and discernment at that time. On the one hand, there is a sympathetic appreciation of their cultural qualities and the failure to study fully in Tasmania a people living as did Europeans perhaps several hundred thousand years previously is deeply regretted. On the other, there is a fair-minded recognition that the decimation of the mainland Aborigines, and even the extermination of the Tasmanians, was mainly involuntary and an inevitable natural consequence of contact through disease and desperation. He claims that both peoples had occupied their lands for many tens of thousands of years, and denies their degeneration to barbarism in this long isolation, pointing rather to a high ability to exploit and adapt to the resources of the continent, leading to a marked cultural uniformity over the mainland. The Tasmanians differed because of isolation by Bass Strait.

This last question is taken further in his 1925 paper, 'Pleistocene changes of sea level and the distribution of man', which is substantially devoted to Australia. No one at that time saw more clearly than he the implications for human migrations and other biological patterns of repeated connections of Australia to New Guinea and Tasmania because of glacio-eustatic oscillations. He was careful to acknowledge that David (1923) and other Australians such as Taylor were cognisant of much of this, but failed to refer to Noetling's important 1910 paper on Bass Strait in this connection, one of his few lapses in gathering in previous Australian literature relevant to his own studies.

There is thus least novelty in his treatment of Bass Strait. He inferred that the Tasmanians crossed Bass Strait as dry land in a glacial low sea level, not necessarily that of the last glaciation, for the maximum of which he accepted an age of 50,000 to 100,000 years. He argued that they probably moved around the Australian coast of that time and up the coastal rivers on their way to the southeastern corner. The mainland Aborigines did not reach this area till after Bass Strait was recreated after the last glaciation.

His remarks on the northern land bridge were more original and percipient than the views others held at the time he wrote. He appreciated clearly that the deep troughs around the islands of Wallacea between the Great Sunda Islands (Java, Sumatra, Borneo) and New Guinea had remained sea through the glacio-eustatic swings of the Pleistocene. Nevertheless he continued to be perplexed by the greatness of the biotic differences between the Asian and the Australasian realms. Against the common scientific opinion of the times, he speculated that the exposed surfaces of the Sunda and Sahul shelves were dry areas in the glacial periods, not jungle as was generally conceived. Indeed it is only recently that others have adopted this view (e.g. Verstappen 1974). The broad Australia-New Guinea shelf would have been particularly inhospitable to water-loving plants and animals (cf. Walker 1972). Moreover exposure of the shelves would have lasted each time some tens of thousands of years only. Of the animals, only man, the dingo and some rodents made the crossing; 'The short-lived and interrupted bridge [between Asia and Australia] has not the significance ascribed to such a bridge by biogeographers'. Recently the role of Torres Strait for man and the rest of the living world was the burden of a conference which made no reference to Daneš, a disgraceful omission to which the present author contributed (Walker 1972). That Daneš had a lively appreciation of biogeographical problems is also revealed in his brief article (1924c) on the effects of the introduction of rats in 1919 to Lord Howe Island, recognising at the same time its catastrophic nature and its great value as a documented experiment in a certain sense in island biogeography.

Daneš goes on to maintain that the drowning of the shelves in the postglacial rise of sea level explains the scarcity of human remains of Pleistocene age in Australia. Even though they occupied the whole continent, these 'Palaeolithic' hunter-gatherers would have concentrated most of all on the sea coast and the lower river courses where food would have been more plentiful, and these are now on the bottom of the shallow, epicontinental seas where they may have been eroded away or buried beneath fresh sediment. Nevertheless he reflects that big kitchen middens such as he had seen on the New South Wales coast near Ballina could have survived on the continental shelf where they might be sought out with the help of submarines. No doubt Vening Meinesz' submarine geophysical exploits at that time on the Sunda shelf were in his mind. He also made the point that Torres Strait was reoccupied by the sea before more advanced cultures from Asia had penetrated so far southeast. All this makes for startlingly modern reading; yet our archaeologists have remained ignorant of Daneš' ideas, though they have arrived at similar ideas independently and in certain ways have gone beyond them (Jones 1968, 1977a, 1977b; Bowdler 1977).

An Isolated Continent

Daneš' other human geographical paper translated into English (1925b), and parts of his earlier paper in Czech (1910d), discuss the effects of isolation on Australia, another theme renewed of late for this smallest continent almost at the centre of the water hemisphere (e.g. Blainey 1966). Daneš approached the matter with the point of view of Lucien Febvre rather than that of Friedrich Ratzel; for him much of the isolation was of social origin.

Thus, although the drowning of the northern continental shelf isolated the continent after the Aborigines had occupied it, in his view it was physical resistance on their part that subsequently excluded sea-going Papuans and Makassans. More advanced and powerful Asiatic peoples, who also probably knew of Australia for many centuries prior to European 'discovery', found it too unrewarding economically to penetrate. This was also true for the Dutch.

Early British colonisation was slow and partial. The 'natural agents' responsible were the 'long, expensive and troublesome journey, deterring emigrants to whom America was more accessible' and the climatic and soil conditions 'considered to be unsuitable for the more intensive forms of agriculture and thought to allow but extensive stock-farming on a grand scale'. However, 'just as important' were 'artificial ones due to man', including the convict character of the first settlements, the despotic regime which would seem uninviting to free settlers, and the discouragement of other than British immigrants.

The inrush of gold-diggers in the 1850s, not all of them British, added weight and wealth to the demand for colonial autonomy. But 'the squatters took hold of the new institutions and only after long political struggles' were the newcomers able to reform the parliamentary systems. Autonomy led to bitter antagonisms between the colonies, which had 'deplorable results' such as different railway gauges. These arose despite common elements such as 'English social customs', the 'dislike of foreigners' and the 'racial aversion' for coloured people. The miners' unions did, however, bring a unity of thought transcending colonial boundaries.

Once the gold-fever subsided, white immigration slumped once more. Many of the deterrents Daneš had enumerated still operated and 'The English reserve, the selfconsciousness of the Sons of Albion delighting in their splendid isolation from the influences of the European mainland, continued in the Australians in a new form'. Moreover conservative elements, especially the squatters, were hostile to further white immigration because they were 'struggling hard against wide circles of people getting hold of the soil, the liberal professions and securing civil rights'. In contrast, propertied classes promoted the import of coloured labour such as kanakas in Queensland.

Perhaps it needed a complete stranger to give in a view of nineteenth century Australia such emphasis to the differences that arose within and between the colonies despite the common cultural tradition and antipathy of most of the settlers for anyone not of British stock. However, characterisation of events as a struggle between the squatters and other settlers over-simplifies issues and transactions. Reference to Coghlan's massive *Labour and Industry in Australia* (1918) would have counteracted the polemical writings of J.D. Lang which he does cite. Failure of private enterprise to build railways in Australia resulted not only from lack of interest by conservative, wealthy elements and hostility to foreign capital but even more from 'lack of entrepreneurial ability' (Butlin 1959). Government-run railways were profitable despite the thin spread of productive resources compared, for example, with the U.S.A. where the railways were in the private sector.

The unification of the Australian colonies into a Commonwealth brought a whole continent into a single political unit and accentuated the urge for the country to go its own independent way in the world, paradoxically 'backed by the powerful though invisible support of the British Empire with its fleet'. For Daneš, this was the work of the 'progressive elements and especially of those who were politically organised in the Australian Labour Party', for whom 'each colony was a stronghold of conservative elements represented by the great land-owners'. Fear, particularly on the part of the workers, of an influx of Asians, also contributed to the development of national feeling. Again he over-simplifies a complex process to bolster his theme of isolationist influences. Several elements in Australian society — not only organised labour — were beginning to see virtues in a federation. Associations of manufacturers were pressing for a federal system since small-scale production for separate colonial markets made competition with imports impossible, and another influence — counter to Daneš' thesis — was the growing recognition in the 1880s and 1890s that to sell meat and dairy produce in Britain all the colonies would have to adopt the same standards of production and marketing (Linge 1979). Moreover the notion of the 'yellow peril' was more broadly present through the community than he represents. Indeed the contagion seems to have afflicted Daneš himself, leading him to neglect other fears of external danger, namely that of a Russian invasion.

Daneš also held that labour was chiefly responsible for the White Australia policy whereby the fundamental apartness of the continent would be accentuated by making it the sole inheritance of the British people. However, his most detailed criticism was reserved for the adoption of a common tariff system to protect Australian manufacturing (claiming, mistakenly, that only Victoria had such a tariff system). Insulation against international competition leads to sloppy workmanship, inferior goods, lack of initiative and high prices. Behind a wall of high duties 'labour is levelled down, there is neither inducement nor wish to specialize by practice or to excel in workmanship, and so the spirit of enterprise and ambition is nipped in the bud'. He also canvassed another argument familiar today that the privileges afforded the manufacturing sector impose a burden on rural producers. The small man in the country has to pay for the easy life of organised labour in the big cities with their conveniences and entertainments.

After deploring the effects of isolation on intellectual and aesthetic life (science alone escaping, 'its own reward in itself'), he concludes that the man-made component of isolation will have to be curtailed and Australia will have to make efforts 'to extricate itself from the critical situation brought about by too one-sided an exploitation of its island position and of its natural resources in carrying through social reforms for which the humanity [*sic*], still regarding systematic work as an evil, is not yet ripe'. Today's newspaper editorials echo many of his arguments. Indeed his writings in human geography are as penetrating as those in physical geography and deserved an attention they never received. Had they been known, some of his insights, deriving from an ardent Czech nationalism at once both ancient and very young, could scarcely have escaped reference in modern works such as Blainey (1966) and Spate (1968).

Conclusion

From Daneš' energetic first visit of eight months, there appeared substantial writings about Australian physiography and his long stay of two years as consul resulted in matching efforts, appropriately in human geography. Some of the dangers facing a visiting foreign academic he largely overcame as, for instance, in his command of earlier Australian literature relevant to his own studies. Likewise he established good communication with his contemporaries active here. In practical ways he adapted readily to the 'bush' and for the most part he avoided the risks of misinterpreting strange things in a strange land through the blinkers of different language and culture.

On the debit side it has to be admitted that much of his writing was overlooked in the country to which it was devoted, though the incidence of neglect has been markedly variable and is by no means solely governed by place and language of publication. Daneš indeed tried hard to obviate this by publishing much in English, French and German instead of his native tongue and actually in Australian journals. His karst work was long disregarded, partly because karst was itself neglected here for a long time. However, his contributions to human geography were in fact hidden from Australian eyes and still are to a large extent.

In addition some of his work was misunderstood in Europe because of real differences in the nature of the two continents and to some degree because of the Procrustean bed of terminology. This was no doubt inevitable for a man who ventured so far afield.

Daneš' Australian contribution can be evaluated in two ways — by comparison with contemporary writings so estimating its place in the history of Australian geography, and by relating his ideas to present-day views and controversies. On the first count it is plainly evident that historically he deserves a place of some importance, one which has not yet received any semblance of recognition (e.g. Scott 1977). As to the second it is true that many of his views no longer carry conviction. But time bowls over most scientific edifices and this process has certainly not affected Daneš any more than his most illustrious Australian contemporaries in geography and geology. On the positive side some of his interpretations still need consideration and testing in relation to current discussion. Without doubt, he took advantage of the special opportunities afforded the scientist from one of the world centres of academic effort in a largely virgin study area and he brought back, in the best sense of the word, 'spoil of the old world's farthest end'.

Acknowledgments

Many years ago Professor Josef Kunský of Prague kindly responded to my inquiries about Daneš by sending me the extremely full obituary notices and bibliography of his writings from the *Sbornik Československy společnosti zemědvědné v Praze.* I greatly regret that this modest effort to explore the Australian role of his illustrious compatriot did not come to fruition before his own death. Recently when the occasion of this book, for which it seemed an appropriate component, made it urgent to secure further materials from Czechoslovakia, Dr Jan Jelinek and the Anthropos Museum in Brno helpfully came to the rescue. These writings in Czech were a black box for me but Mr Henry Polach of The Australian National University generously spent much precious time to give me a good inkling of their contents.

References

Andrews, E.C. 1910. Geographical unity of E. Australia in late and post Tertiary time, with applications to biological problems. *Journal and Proceedings of the Royal Society of New South Wales* 44: 420-80.

Blainey, G. 1966. *The Tyranny of Distance: How Distance Shaped Australia's History*. Sun Books, Melbourne.

Bowdler, S. 1977. The coastal colonisation of Tasmania. Pp. 205-46 in J. Allen, J. Golson and R. Jones (eds.), *Sunda and Sahul Prehistoric Studies in Southeast Asia, Melanesia and Australia*. Academic Press, London.

Bowler, J.M. *et al.*, 1976. Late Quaternary climates of Australia and New Guinea. *Quaternary Research* 6: 359-94.

Brandt, B. 1928. Jiří V. Daneš. *Petermann's Geographische Mitteilungen* 74: 231.

Brook, G.A. and Ford, D.C. 1976. The Nahanni North Karst: a question mark on the validity of the morphoclimatic concept of karst development. *International Speleology 1973* 2: 43-57.

Browne, W.B. 1934. Some peculiarities in the drainage system of the Australian continent. *Australian Geographer* 2(4): 13-19.

Butlin, N.G. 1959. Colonial socialism in Australia, 1860-1900. Pp. 26-78 in H.G.J. Aitken (ed.), *The State and Economic Growth.* Social Science Research Council, New York.

Coghlan, T.A. 1918. *Labour and Industry in Australia.* Oxford University Press, Oxford.

Daneš, J.V. 1910a. Some problems of Queensland hydrography. *Queensland Geographical Journal* 25: 75-81.

Daneš, J.V. 1910b. Report of a tour along the Dividing Range (better Plateau) from Aramac to Pentland. *Queensland Geographical Journal* 25: 83-101.

Daneš, J.V. 1910c. Physiography of some limestone areas in Queensland. *Proceedings of the Royal Society of Queensland* 23: 45-86.

Daneš, J.V. 1910d. Geografické a politickohospodářske problémy Australiského soustati [Geographical and economic problems of the Australian Commonwealth]. *Sbornik české společnosti zeměvědné v Praze* 16: 113-34.

Daneš, J.V. 1911. On the physiography of north-eastern Australia. *Vestnik Královské české společnosti nauk* 32:1-18.

Daneš, J.V. 1912a. La capture de la haute Flinders. *La Géographie* 26: 263-9.

Daneš, J.V. 1912b. La région des rivières Barron et Russell (Queensland). *Annales de Géographie* 21: 246-63.

Daneš, J.V. 1912c. Krajinné typy v severorých Australii [Natural regions of northeast Australia]. *Sbornik české společnosti zeměvědné v Praze* 18: 214-21.

Daneš, J.V. 1913. Die tropische Nordostküste Australiens. *Die Erde, Weimar* 1: 346-63.

Daneš J.V. 1916 Karststudien in Australien. *Vestnik Královske české společnosti nauk* 7: 1-75.

Daneš, J.V. 1924a. Limestone physiography in Australia. Pp. 337-40 in *Zbornik vadova posvećen Jovanu Cvijiću: Receuil de travaux offert à M. Jovan Cvijić.* Belgrade.

Daneš, J.V. 1924b. *Původ a Zanikání Domorodců v Australii a Oceanii* [Origin and Extinction of the Aborigines in Australia and Oceania]. Nakladem České Gratické Unie, Prague.

Daneš, J.V. 1924c. Pohroma na ostrové Lord Howea. *Širým svetem, Prague* 1: 7-10.

Daneš, J.V. 1925a. Pleistocene changes of sea level and the distribution of man. *Scottish Geographical Magazine* 41: 289-98.

Daneš, J.V. 1925b. The isolation of Australia. *Národopisný věstnik československý* 18: 175-90.

Darwin, C. 1845. *Journal of Researches into the Natural History and Geology of the Countries visited during the Voyage round the World of H.M.S. Beagle. Under command of Captain Fitz Roy. R.N.* Ward Lock, London.

David, T.W.E. 1923. Geological evidence of the antiquity of man in the Commonwealth, with special reference to the Tasmanian aborigines. *Papers and Proceedings of the Royal Society of Tasmania* 1923: 109-50.

David, T.W.E. 1950. Edited by Browne, W.R., *The Geology of the Commonwealth of Australia.* Arnold, London.

Dvorský, V. (ed.) 1928. Life and work of Dr. J.V. Daneš. *Sbornik české společnosti zemépisné v Praze* 34: 205-26.

Ford, T.D. 1978. Chillagoe — a tower karst in decay. *Transactions of the British Cave Research Association* 5: 61-84.

Hamilton-Smith, E. 1966. Caves of the Chillagoe district, North Queensland. *Helictite* 4: 53-9.

Hedley, C. 1911. A study of marginal drainage. *Proceedings of the Linnaean Society of New South Wales* 36: 13-38.

Hedley, C. 1925. A drained river mouth at Cairns. *Transactions of the Royal Geographical Society of Australasia (Queensland)* 1: 69-72.
Hedley, C. and Taylor, T.G. 1907. Coral reefs of the Great Barrier Reef, Queensland: a study of their structure, life-distribution, and relation to mainland physiography. *Report of the Australasian Association for the Advancement of Science* 1907, 397-413.
Jardine, F. 1925a. The physiography of the Port Curtis district. *Transactions of the Royal Geographical Society of Australasia (Queensland)* 1: 73-110.
Jardine, F. 1925b. The drainage of the Atherton Tableland. *Transactions of the Royal Geographical Society of Australasia (Queensland)* 1: 133-48.
Jennings, J.N. 1966. Jiří V. Daneš and the Chillagoe Caves district. *Helictite* 4: 83-7.
Jennings, J.N. 1968. Syngenetic karst in Australia. Pp. 41-108 in P.W. Williams and J.N. Jennings. *Contributions to the Study of Karst.* The Australian National University, Department of Geography Publication G/5.
Jennings, J.N. 1977. Caves around Canberra. Pp. 379-85 in *Proceedings Eleventh Biennial Conference Canberra 1976.* Australian Speleological Federation, Sydney.
Jennings, J.N. and Sweeting, M.M. 1963. The Limestone Ranges of the Fitzroy Basin Western Australia. *Bonner Geographische Abhandlungen* 32: 1-60.
Jones, R. 1968. The geographical background to the arrival of man in Australia and Tasmania. *Archaeology and Physical Anthropology in Oceania* 3: 186-215.
Jones, R. 1977a. Sunda and Sahul: an introduction. Pp. 1-9 in J. Allen, J. Golson and R. Jones (eds.), *Sunda and Sahul Prehistoric Studies in Southeast Asia, Melanesia and Australia.* Academic Press, London.
Jones, R. 1977b. Man as an element of a continental fauna: the case of the sundering of the Bassian bridge. Pp. 317-86 in J. Allen, J. Golson and R. Jones (eds.), *Sunda and Sahul Prehistoric Studies in Southeast Asia, Melanesia and Australia.* Academic Press, London.
Lehmann, H. 1954. Das Karstphänomen in den verschiedenen Klimazonen. *Erdkunde* 8: 112-220.
Lehmann, H. 1960. La terminologie classique du karst sous l'aspect critique de la morphologie climatique moderne. *Revue de Géographie de Lyon* 1960, 1-6.
Linge, G.J.R. 1979. *Industrial Awakening: A Geography of Australian Manufacturing 1788 to 1890.* Australian National University Press, Canberra.
McAuley, J. 1946. *Under Aldebaran.* Melbourne University Press, Melbourne.
Maksimovich, G.A. 1962. Karst Australii [Karst in Australia]. *Gidrogeologiya i Karstovedeniye* 1: 153-71.
Marker, M. 1977. A geomorphological assessment of the Chillagoe karst belt, Queensland. *Helictite* 14: 31-49.
Marks, E.O. 1913. Notes on part of the Burdekin valley. *Proceedings of the Royal Society of Queensland* 24: 93-102.
Marks, E.O. 1924. Some doubts in Queensland physiography. *Proceedings of the Royal Society of Queensland* 30: 1-18.
Marks, E.O. 1931. The physiographical significance and non-migration of divides. *Proceedings of the Royal Society of Queensland* 43: 52-61.
Noetling, F. 1910. The antiquity of man in Australia. *Proceedings of the Royal Society of Tasmania* 1910: 231-61.
Ollier, C.D. 1978. Tectonics and geomorphology of the Eastern Highlands. Pp. 5-47 in J.L. Davies and M.A.J. Williams (eds.), *Landform Evolution in Australasia.* Australian National University Press, Canberra.
Poole, W. 1910. Physiography of N. Queensland. *Report of the Australasian Association for the Advancement of Science, Brisbane 1909:* 316-18.

Robinson, T. 1978. A question of age. *Tower Karst Occasional Paper* 2: 18-36.
Scott, H.I. 1977. *The Development of Landform Studies in Australia.* Bellbird, Artarmon.
Spate, O.H.K. 1968. *Australia.* Benn, London.
Stanley, G.A.V. 1927. The physiography of the Bowen district and of the northern islands of the Cumberland Group (Whitsunday Passage). *Report Great Barrier Reef Committee* 2: 1-51.
Süssmilch, C.A. 1938. The geomorphology of eastern Queensland. *Report Great Barrier Reef Committee* 4: 105-34.
Sweeting, M.M. 1973. Karst landforms and limestones. *South African Geographical Journal* 55: 81-8.
Taylor, T.G. 1911. Physiography of eastern Australia. *Commonwealth Bureau of Meteorology Bulletin* 11.
Verstappen, H.Th. 1974. On palaeo climates and landform development in Malesia. Pp. 3-35 in *Modern Quaternary Research in South East Asia.* Balkema, Rotterdam.
Walker, D. (ed.) 1972. *Bridge and Barrier: the Natural and Cultural History of Torres Strait.* The Australian National University, Department of Biogeography and Geomorphology Publication BG/3.
Williams, P.W. 1978. Interpretations of Australasian karsts. Pp. 259-86 in J.L. Davies and M.A.J. Williams (eds.), *Landform Evolution in Australasia.* Australian National University Press, Canberra.
Wilson, P. 1975. Observations of the geomorphology of the Chillagoe limestones. Pp. 69-81 in *Proceedings Tenth Biennial Conference Brisbane 1974:* 69-81. Australian Speleological Federation, Sydney.
Young, R.W. 1977. Landscape development in the Shoalhaven River catchment of southeastern New South Wales. *Zeitschrift für Geomorphologie* 21: 262-83.

Diverse Faces of Historical Geography

Chapter 5

Oral Historical Geography

*Bruce Ryan**

On the title page of *India and Pakistan* there is just one quotation, from Strabo: 'We must hear accounts of India with indulgence, for not only is it very far away, but even those who have seen it saw only some parts of it, and most of what they tell us is from hearsay' (Spate 1954). The book that follows, of course, is anything but hearsay. Heresy in parts, perhaps, but certainly not hearsay. Its documentary scholarship carries the conviction that comes only from long and perceptive familiarity with the land itself. By contrast, hearsay includes gossip, rumour, anecdote, myth and legend — if that is the correct declension — as well as the oral traditions of many human groups, passed down from generation to generation. As a valid source of historical information or 'evidence', it tends to be dismissed, often superciliously, by scholars whose graduate training conditioned them to think that knowledge and certainty reside only in documents, and that the grander the documentary 'truth', the grander and more venerable will be the surroundings in which it reposes. To understand the force of this supposition, one need only compare the National Archives in Washington, D.C., which entomb the Declaration of Independence, with Jefferson's desk at Monticello, on which it was written.

* *K.B. Ryan is Professor of Geography and Department Head at The University of Cincinnati, Ohio, U.S.A. After taking B.A. and M.A. degrees in Geography at The University of Sydney, he went on to The Australian National University for his doctorate, writing a thesis under the baleful eye of O.H.K.S. He graduated in 1966.*

Yet we may doubt whether Spate would discount oral evidence quite so completely as most historians are wont to. In his writings, explicit examples of 'oral historical geography' are hard to find, despite the approval implied in the quotation from Strabo. Much of that writing is nonetheless quite sympathetic to the oral recollections of eye-witnesses. 'On being an expert' (Spate 1961) tells of a young academic researcher, timorously venturing into some foreign field area, wondering how to incorporate 'tantalizing scraps of really confidential stuff' into his report, asking himself how to size up informants, trying to ingratiate himself locally by being 'coarsely cynical in club or pub', but leaving out of his report 'a lot of stories which will go down very well in the Common Room', once he returns from the field. Spate very clearly understands the context of gathering oral testimony. In 'Manuel Godinho de Erédia' (Spate 1957), he goes beyond mere understanding. The opening pages are written — as only Spate could — in the 'historical present' of A.D. 1545, and even include 'fictitious' passages of direct speech, based on no written record whatsoever but crisp with authenticity. On Mbananakoro, an imaginary (?) South Pacific island dependency, Spate (1965) tells how the authorities tried to reconstruct traditional choreographies for potential tourist consumption, but diluted the traditional dances (as oral history sometimes does) by toning down what might offend visitors from 'Shaker Heights, Ohio, or Virginville, Pa.' — in this case, 'the more overt expressions of the procreative instinct'.

In all these examples, Spate reveals insight into the perils and promise of gathering oral evidence in the field, and preparing it for polite consumption. His stance is a humanistic one, full of concern that the researcher translate and transmit accurately what he gathers, while not compromising his informants in their own communities. His command of scholarly English is matched by an equal facility with the vernacular of illiterate informants who express themselves, he says, in 'the halting language of earthy experience, articulate only in concrete fact or a dry salty peasant image or a laconic, gnomic summation of the eternal commonplaces' (Spate 1961).

Talking History

Yet Spate's approval of a few mavericks in pursuit of 'oral historical geography' does not establish a field of scholarship. If such a hole does exist in the mesh of research, we must decide whose darning needles should repair it. At the moment oral historians have the best selection of needles, and all the prizes for darning. Still slightly faddish, but certainly the most respectable direction oral investigation has so far taken, oral history offers to geographers the most illuminating insight into where oral studies might take them. It is a robust, almost belligerent field. Its practitioners are brash young men and women tramping the back-streets and byways with their tape-recorders, fired with an urgency to rescue from extinction the unwritten recollections of lives already largely lived. Few if any of them conform to the stereotyped historian patiently plodding through musty archives.

The Messiah of the 'oral history movement' was the late Professor Allan Nevins of Columbia University, whose litany of achievements now appears in the preface of every oral history directory, review and textbook. It is not unlike the Apostles' Creed, although those of us who remember what Spate did to Toynbee, another Messiah, will step rather gingerly into this arena. Columbia University's first interview was conducted on 18 May 1948, at 120 East 75th Street, New York City (Starr 1971: 278), an address which will no doubt qualify in 1998, after the statutory fifty years have elapsed, for inclusion in the National Register of Historic Places. For a decade, historians largely ignored the 'oral movement' within their ranks, the proselytes meekly fortifying themselves with the hope that future historians would harvest whole libraries of monographs from the narratives collected in their cassettes. Oral historians advertised their services and the urgency of their mission. They argued that the 'confidential letter' was obsolete, that statesmen no longer kept diaries, that business magnates could 'ghost' their autobiographies (and guarantee themselves a place in the starry firmament) if only they would submit themselves to a candid oral interview. Politicians were especially receptive, although the wheel did turn, and the overt glorification of a self-styled 'élite' was elbowed aside to make room for some 'history from the ground up' (Benison 1965: 75). When Dorson (1964) chastised his fellow American historians for rejecting 'folk and other oral traditions' as adamantly as they had, he was able to identify five significant areas of research in which oral evidence had much to contribute: popular attitudes, prejudices and stereotypes — as manifested geographically, for example, in regional slurs ('keep

Wales tidy; dump your garbage in England'); national myths, symbols and images — such as those identified in Lowenthal's 'The bicentennial landscape: a mirror held up to the past' (1977); the disentangling of 'fact from fancy', as the folklorists attempted in their Motif K.185.1, 'Deceptive land purchase: ox-hide measure'; the corroborative verification of incidents; and the provision of data on minority groups, on 'tradition-directed' social enclaves that preserve an older, even archaic culture often lacking its own written records.

After lurking for a decade on the fringes of academic respectability, oral history came into its own. Tape-recorded narratives began to find their way into books. The 'buoyant quality' of 'creative pioneering' began to pay off (Starr 1971: 283). In 1967, the Oral History Association was organised. Not ten years later, its membership exceeded 1,000 (Meckler and McMullin 1975: 1), compared with 7,000 members recruited by the Association of American Geographers after a campaign that began in 1904. By the 1970s, in what will surely prove to be an important diffusion study, forty coherent 'programs' in oral history had been mounted by American universities, and the crusade launched by Nevins had captured a million pages of transcript in its attempt 'to catalogue and preserve the envelope of sound in which we all live' (Benison 1965: 76). Some said that the 'movement' had revitalised the teaching of history, leading students gently out of their stereophonic adolescence, preparing them for a job market (not overly sympathetic to historians) in such labour-deficient fields as records management and public relations, offering them every incentive to absorb the necessary documentary background before venturing out to interview 'old men drooling about their youth' (Storm-Clark 1978: 190). History became an exciting, audio-visual, computerised, urgent, collaborative enterprise. The dizzy expansion of data soon outpaced their own dissemination and use.

Talking Geography

At least some of this oral testimony described past geographical patterns. It trespassed without apology or acknowledgment upon the turf of the historical geographer. Few geographers noticed the incursion. Baker's *Progress in Historical Geography* indexed 'oral tradition' only once, and that in reference to tribal Africa (Baker 1972: 204). The recently launched *Journal of Historical Geography* has yet to publish research that relies primarily on oral testimony, although the *American Anthropologist* is full of it, albeit with respect to social organisation rather than 'territory'. Admittedly, little of direct value to the historical geographer is to be found among the categories into which Vansina's authoritative typology divides the range of 'oral traditions' (Vansina 1965: 144; 1971: 451). Memorised lists of place-names do quite often recount the chronological succession of places along a tribe's migration route and may even imply something of its past demographic structure. Aetiological myths are those that 'explain' the origin of cultural traits and natural features without recourse to divine intervention. Some of them, viewed through a glass darkly, to be sure, hint at mythical landscapes and environments, as do certain Icelandic legends invoking vulcanism. Then there are lists of personal names, many of them royal genealogies recited from memory at funerals or coronations, which sometimes enable scholars to 'mesh time and place', as the saying goes. Among the orally transmitted 'commentaries' that control social procedures, Vansina includes 'legal precedents' that may determine, for example, the origins of property boundaries and land ownership. But these are examples of only four out of the twenty-two types of oral tradition that he categorises: there is very little of geographical significance in the others — in the slogans, titles, liturgical or panegyric poetry, familial narratives and so on. Such evidence suffers far too routinely from progressive 'structural amnesia', as the anthropologists call the process whereby traditions shed their practical purposes and slide into oblivion.

That said against 'folk memories', however, there remains much of crucial utility for the historical geographer in other forms of oral evidence. Not without cause do the anthropologists separate *personal* memories from all that are not. The protean residues they lump together as 'myth'. Sometimes, however, oral evidence can help dispel some current 'myth' enshrined in documentary or field evidence. One hears of geographers mapping urban functions who uncritically copy into their field-books every faded store sign in sight, never bothering to ask local residents whether a self-proclaimed hardware store might not really be a 'front' for an unofficial gambling casino. At least one undergraduate thesis on Alice Springs fell into just such errors. Even official records tend to be quite selective in their coverage. The present renovators of one old office structure in downtown Cincinnati, long vacant, were surprised to find opening off an upstairs corridor what looked like hotel rooms, except for a series of 'viewing' windows alongside each door — through which, it dawned on them, a selection of female pulchritude was once appraised by potential male customers. The building's plans and courthouse records were strangely silent on the matter. In such cases, we can only acknowledge the truth in footnotes that say 'personal communication'.

Deficiencies in the documentary record are legion, even in as consciously, scrupulously and recently recorded a process as the settlement of New Zealand, which Powell (1972: 73) has contrasted with the contemporaneous British muddle, where self-serving lawyers contrived to keep land dealings 'circuitous, involved and secret'. Torrens title — that South Australian miracle of administrative tidiness — merely simplified and codified the relationship of two variables (the ownership and location of land) out of the thousands that comprise the transaction we call 'settlement'. No colonial 'lands and survey' office could possibly have anticipated the breadth of interest in local history today or the demand for records that it arouses. As Jack (1977: 25,29) has reminded us, that interest runs all the way from the archaeological excavation of junk yards and bottle-dumps to the reminiscences of 'people steeped in the locality and its affairs, often without formal training, who talk on very specific local topics, a single industry or a single house or a single family'. Nor is local research helped by the often 'excessive centralisation of the archives service' in remote cities where libraries exclude inquirers because their finger-nails are dirty and their education incomplete.

Oral historical geography not only repairs some of these deficiencies in the documentary record, but often conveys a greater variety of past 'attitudes' towards geographical space than the written record. Only certain kinds of informants conceptualise 'space' as the physical or social scientist does. Not everyone knows or admits that certain places, for certain people, have had sacred or magical connotations. Not everyone is able to spend their sophisticated, aesthetically-sensitive lives and imaginations ranging over vast areas, roaming the world, trying to differentiate each kind of area from all others and ending with a large but fragmented conception of geographical space. Others are confined in their experience to very circumscribed areas; they view space with the eyes of children, or take the 'mythical-magical' view, or perceive it in the age-old traditional constructs of the 'primitive' society to which they belong (Sack 1978). Sack calls this the 'unsophisticated-fused' view of geographical space. Since it tends to evoke oral rather than written transmission of the past, historical geographers should willingly confront the 'sonic environment' that contains it, with their tape-recorders in the field.

Exhortation is one thing; successful practice quite another. Fortunately, historical geographers have already compared and combined oral and written evidence successfully, though not very often. McCourt (1972), for example, has done so in studying the Irish open field (or rundale) system and the clustered, 'clachan' type of settlement associated with it, both of which were swept away in what he calls 'the seemingly silent social revolution' between 1750 and 1850. The demise is poorly documented, but McCourt found what was available for North Ulster. He also sought out those vestigial townlands which still possessed a 'social environment harbouring traditions about pre-enclosure times and practices' — places where winter fireside gatherings (*ceilidhe*), work boons and wakes were still the vehicles for oral transmission, although open-field farming and the Gaelic language were no longer parts of that residue. When mapped, the pattern formed by these isolated townlands was more restricted in area than the regions for which documentary ('historical') evidence of rundale had been obtained. Much of it was upland. Unexpectedly, however, the two patterns were only partly congruent: about one-sixth of the townlands with oral traditions of rundale lay completely outside the area for which there was historical evidence of open fields and clachans. If oral evidence can thus enlarge the geographical coverage of written evidence, it is surely worth pursuing.

McCourt's study is also a timely rebuke to those who would separate field research from archival research. We should not need reminding of the fruitful interplay between the two. Spate concluded 'Bush and city' by relating how Edward Gibbon caught the notion of writing the 'decline and fall' of Rome while musing in its ruins (Spate 1956: 184), and as citation-dense a work as Clark's *History of Australia* very candidly conceded 'the power of place', which Spate had called 'the province of geography'. His most fruitful sources of all, said Clark (1973: 470), were two not listed in his footnotes.

One was to stand at the South Head of Sydney Harbour and

gaze at what Magellan called that 'very vast sea', and then gaze at Sydney Harbour. That was always exhilarating and always rewarding because it turned the mind towards what this country did to those who made the long journey over the oceans. A regular visit to the Henry Lawson statue in the Sydney Domain, on the other hand, turned the mind towards what men have done to each other in Australia.

Talking Discredited

Is all this too extravagant an advocacy of oral historical geography? If so, the next step is to argue as strenuously as possible against its pursuit. Such a critique will come from three directions: it will ask whether oral evidence can be validated, verified or refined to the satisfaction of sceptical scholars; it will question the usefulness of conceptualising the past in a mixture of 'mythical' and 'factual' terms; and it will expose many problems with the technique as such — with collecting, translating and publishing oral information.

Oral evidence is readily distorted, deliberately or otherwise, although the purveyors of 'professional' history, according to Dorson (1964: 227), are as prone to ethnocentrism as the purveyors of folklore. Temporary divisions within a single society may also toss out radically different versions — and ultimately 'traditions' — of the same events. During the nineteenth century, for example, official and unofficial attitudes towards Australian bushrangers had little in common (Walker 1964). Respectable parliamentarians reviled them as predators and outlaws, as immoral and vagrant as cats. Yet bush-workers, emancipists and miners often sympathised with these 'wild colonial boys', and by the end of the century had glorified their defiance of authority in the bush ballads, and captured their spirit in the nationalism that culminated in a federal government in 1901. Informants of one race may also lie to questioners of another race, whom they distrust or detest. The prospect is especially likely when indigenes are faced with police prosecution by 'colonial' overlords, and conflicting value systems are not offset by reciprocal respect and personal confidence (Strehlow 1936). Australian Aborigines have been embittered by the cavalier irresponsibility of scientists to whom they had given sacred male objects and access to sacred male ceremonies, who later revealed these ancient tribal secrets to white women and scoffing museum visitors. Subsequent oral testimony therefore tended to be distorted or incomplete.

In many societies, the mythical past mingles invisibly with the 'true' historical past, until neither is unequivocally recognisable. In 1970, however, one seminar devoted to 'Folklore and Traditional History' did conclude — from the evidence of modern Africa, ancient Greece, medieval Iceland, early national Sweden and the feudal Highlands of Scotland — that the historical content of their findings (and the historical attitudes of 'tradition bearers') more than neutralised the leavening of folklore (Dorson 1971: 81). Indeed, the histories of African countries are now being *written* in large measure from *oral* traditional sources. Even so, the exact relationship between oral evidence and oral tradition is very hard to establish: first, because 'one man's history is another's fable'; second, because 'oral sources are people; in a short time they die, and there is no dependable way of consulting them once they are dead' (Bynum 1971: 83).

Especially intriguing to the historical geographer is the frequent association of mythical events with specific locations. Harwood (1976) has attributed this to the inability of many non-literate societies to record 'time depth' chronologically. Instead, a series of *locations* is substituted, each location being a mnemonic device for recalling a particular mythical event in the correct sequence. Recalling Malinowski, she regards myths as 'institutional charters' which tether such 'institutions' as fishing, canoe building or inter-island trade to certain parts of the environment. Myths are a ritual re-enactment, in correct logical precedence, of how these ecological relationships originated — analogous, she claims, to the stations of the cross. Her interpretation, however, comes fifty years after Malinowski's field-work in the Trobriands, on which it is based. That is perhaps a little too belated, and too subtle by half, for the rampant oral historian of today.

Oral evidence suffers from other limitations in non-literate or pre-literate societies. As early as 1915, Lowie fired a salvo of objections at those prepared to accept such evidence on its face value. He claimed that myths were not stable over time, that oral testimony all too often assigned the most unwarranted significance to trivial events, that many Amerindian tribes could not even relate how their immediate ancestors obtained the horse, that corroboration of myth by a neighbouring tribe indicated the diffusion of folklore, not the verification of fact.

> If we do not accept aboriginal pathology as contributions to *our* pathology, if we do not accept aboriginal astronomy, biology, or physics, why should we place primitive history alone on a quite exceptional pedestal, and exalt it to a rank co-ordinate with that of our own historical science? (Lowie 1917: 162.)

In similar vein, Law (1969: 235-6) also gave much more credence to field evidence than oral tradition in tracing the pre-European cultivation of kumara in the South Island of New Zealand. At best, oral testimony merely offered clues as to places where better, more tangible evidence might be worth seeking. Today, courts of law are just as parsimonious: the Indian Land Claims Commission of the United States Department of Justice looks with dubious eyes on those Indian claimants whose argument rests almost exclusively on father to son oral tradition (Dorson 1971: 79). Bank managers also prefer written IOUs to telephoned assurances, however ardently pressed.

Yet disentangling historical fact from historical myth is scarcely more precarious than trying to separate contemporary facts from contemporary myths, as Lowenthal (1975) has insisted. The present landscape is thick with alterations that 'make history conform with memory', and assuage the appetite for nostalgia, continuity and permanence. The Danes have even devised a 'nostalgia index' for their 'historical landscapes' (Newcomb 1972). The hateful present and the dreadful future — rose-tinted prisms through which the past is recalled — conspire to distort oral testimony. So do 'selective recall', which suppresses painful memories and tidies up the remembered geography, and 'historical myopia', which may perpetrate frauds, forgeries and fictions such as Piltdown Man, the Cardiff Giant or Archbishop Ussher's geochronology.

The fluxion of language itself further reduces the precision of oral recall. Semantic change, word by word, is problem enough, but whole vivid chapters of the past are lost with the passing of community patois. Locational shifts in linguistic boundaries can destroy, once and forever, some ancient intimacy between language and landscape. In Wales, for example, the population capable of communicating in Welsh has been reduced since 1961 by suburbanisation and the incursion of English-speaking tourists and retirement communities (Bowen and Carter 1974, 1975). Had English television been primarily responsible, as true patriots complained, decline would have occurred everywhere, which it has not. The loss of Welsh was most apparent in those communities where the proportion able to write the language was significantly smaller than the proportion able to speak it (Bowen and Carter 1975: 10). This implies that oral historical geographers should collaborate promptly with their archival brethren in studying these disjunct communities, rather than more durable linguistic enclaves.

Oral information is typically collected from a single community, family or individual. Because nation-wide oral surveys are rarely attempted, statistical or documentary records are often thought to be essential for research of any very considerable geographical scope. This is not inevitably so. The forthcoming *Dictionary of American Regional English,* now fifteen years in preparation, has pursued a much grander design (Cassidy 1977). Questionnaires, each containing 1,840 questions, were administered in a thousand 'representative' communities selected from the 1960 census, to ascertain the 'traditional vocabulary' of each. In every community chosen, respondents were asked, for example, 'what do people here call "the harness strap that goes under the horse's stomach" ', or 'what do people here call "the amount of wood a person can carry in one trip from the woodpile to the house" '? The first answer is a cinch. The second does vary regionally: an 'armful' in New England, a 'turn of wood' in Texas and the Southeast, and an 'armload' elsewhere. In compiling these data, Cassidy's D.A.R.E. team ignored places that were full of recent immigrants, choosing only 'stable communities'. Finding such communities was a headache; finding them in Florida was a migraine. Informants also had to be native-born Americans who had never moved away from their birthplaces, and whose parents and grandparents had also lived there. No English teachers were consulted, since 'natural speech' was obligatory. Elderly residents were preferred, partly because they always knew 'what the younger people say nowadays'. When aggregated and displayed on computer maps, the five million items of data so far collected begin to disclose an historical cultural geography of the United States, based on oral testimony. In short, the D.A.R.E. project eloquently refutes the criticism that oral interviews are useful only in locally circumscribed studies.

Those who dispute the value of historical geography orally obtained will next attack the techniques employed. They will peddle the dubious notion that 'technical' history frees its practitioners from the emotion and prejudice necessarily inherent in the 'eyewitness' history of the 'participant-observer' (Schlesinger 1971: 339-58). They will require verification of every link in the chain of testimonies, and a comparison of each link with every other for traces of distortion and bias, linguistic or otherwise (Vansina 1965), when they would never think to question the taxonomy or accuracy of a census tabulation. They will take Benison's (1965: 74) estimates — an oral historian's full-time interview program of 100 to 125 hours per year, producing between 2,500 and 2,800 pages of transcript — and ask whether equivalent investments of time in the documentary archives will not yield a higher dividend, in cost-benefit terms of productivity and efficiency — especially when nine-tenths of the oral testimony cannot be substantiated. Finally, they will shake their learned heads at the legal constraints on using or publishing material obtained orally, and often confidentially, in the very recent past (Starr 1971: 287-99). Who may legally seek access to 'open' memoirs? Whose is the copyright of verbatim material at large in the 'competitive jungle' of micropublishing? How can scholars publish uninhibited revelations without risking prosecution for libel? These are weighty questions of real consequence to scholarship. Geographers, historians and hybrids will answer them, as usual, by weighing the prospects against the perils.

Memory Maps

What can people remember about previous spatial patterns, ecological relationships or geographical phenomena? All the previous criticism of oral evidence is pointless if the particulars of historical geography can be remembered with no more certainty than life in the womb. Fortunately, the research literatures of several fields preserve for absent-minded and forgetful scholars what is known about human memory in all its guises.

North American evidence indicates that the chain of oral tradition frequently extends for centuries into the past. The Paiute Indians of Utah, for example, have preserved and handed down historical information about their former neighbours, the Puebloid ('Mukwitch') people, ever since the latter left the region about A.D.1150 (Pendergast and Meighan 1959). These 800 years represent thirty generations; Hopi burial customs have been retained orally for just as long. For 200 years, Eskimo oral tradition has conveyed the destruction of Norse settlements in Greenland, although Lowie, that redoubtable unbeliever in oral fairy tales, cites as refutation the Yuchi Indians of Georgia and South Carolina, who claimed to be the original inhabitants of those areas, but were not there to greet or repulse the first Spanish explorers between 1539 and 1567. Within 300 years of that 'migration', they had forgotten their own origins (Lowie 1917: 166).

European oral tradition plumbs the past just as deeply. As Evans (1971, 1976) has argued, the 'work tradition' is especially well preserved, whether since A.D. 1340, from which date unique East Anglian units of agricultural output can be traced, for 300 years among Yorkshire hand-loom weavers, since the systematic destruction of Welsh colliery records in 1947 (when the mines were nationalised), or since time immemorial with respect to the 'lore of the horse'. Other traditions are less permanently embalmed, and are handed down less intact. Harwood (1976: 786) may be right in claiming that, for most oral information, there is a shallow historical depth reaching back only three or four generations, but beyond that, beyond the reach of memory, lies the realm of myth.

For historical geographers, the acid test of memory may well be whether or not an informant can draw a map of some past spatial configuration. Franz Boas asked Eskimos to do this, and Lewthwaite has analysed the remarkable mental map drawn for Cook in 1769 by Tupaia, the Tahitian, who named or depicted some 130 islands strewn across 5,000 km of the Pacific to the east of Fiji, and thus forced the issue of whether 'a non-literate man was fundamentally incapable of projecting his geographical knowledge on a piece of flat paper' (Lewthwaite 1970: 1). Striking anomalies were certainly revealed by Lewthwaite's analysis — storm-warnings for those wanting to render oral history in cartographic form. Whole archipelagos were found to be displaced, relegated to odd corners of the map because, in Hale's opinion of 1846, Society Islanders defined direction according to the wind's destination, not its origin, thus confounding European navigators who supposed wrongly that the 'north wind' blew from the north. On the *Endeavour* voyage of 1768-71, only half of Cook's own charts were orientated with north at the top. Lewthwaite was further perplexed to find that Tupaia's Polynesians literally conferred place-names on islands only when their European interrogators asked what they were. He found that unrecorded phonetic changes within the Tahitian language made it next to impossible to assign a contemporary identity to many islands 'named' by Tupaia in a dialect now lost forever, and that Polynesian chants preserved the names of islands long since vanished under the waves, like Cantref-y-Gwaelod or Atlantis, as a result of marine erosion or volcanic activity. For such reasons, punters give 'memory maps' only very long odds in the field of historical geography.

How much to expect from 'memory maps' might also be gauged from the recent proliferation of 'mental mapping' by geographers, who clearly attach importance and validity to what people can remember about their surroundings. But whereas behaviourists usually want the most uninhibited, untutored 'free recall' attainable by their respondents, historical geographers try to extract the most complete 'verisimilitude'. One group is interested in the composition and arrangement of mental maps, the other in their content; one asks how recall occurs, the other what is actually recalled. Oral historians are similarly schizoid. One recent oral history of Appalachia — though fascinatingly authentic and next-of-kin to oral historical geography (like Hatfield to McCoy) — disgorges its material in an almost Joycean stream-of-consciousness babble (Shackelford and Weinburg 1977). It provides no answer to the methodological question: 'what theoretical notions about geographical memory are confirmed by this study'?

Instead, mental mapping exercises routinely test such concepts. They remind us that what is sketched on paper from memory is only the bare bones, the 'spatial or skeletal framework', of the plump geographic image actually lurking in the respondent's mind. Of this 'latent or potential map' we obtain only a hint, and this, moreover, is constrained or filtered by all sorts of contextual qualities — by the complexity and areal extent of the map coverage itself; by the 'legibility' of certain dominating landmarks, and by the 'site recalcitrance' of vaguer, less 'imageable' places; by 'graphicity', which includes a person's general spatial ability to perceive and retain the structure and proportions of forms in their entirety, as well as the ability to express them cartographically (Pocock 1976).

Environmental psychologists also remind us that, in their experimental research, the terms 'perception' and 'cognition' refer much more commonly to objects than to whole environments. Here, scale is of the essence. As expected, mental maps of neighbourhoods are much more sophisticated and complete than mental maps of a whole city attempted by the same respondents (Ittelson 1973: 3). Nor, apparently, are sex differences in graphicity the illusory prerogative of a 'male chauvinist pig' like my colleague, Wolf Roder (1977). Pocock (1976: 511) found evidence in mental maps of Durham, England, for a 'female predilection for security and containment' that produced a spatial structuring distinctly different from the 'greater rectilinearity' in maps recalled by males. Again, since 'memory maps' are often the work of elderly respondents conjuring up and sketching what impressed them as children or adolescents, the process of maturation — of changes in a person's 'developmental psychology' with age — will surely influence those recollections. It matters considerably, for example, whether a spatial image was formed in the 'topological' space of the pre-school child, with its unabashedly egocentric view of the world, or in the Euclidean space of adolescence and beyond, with its fixed systems of reference (Hart and Moore 1976: 269). The permanence and fidelity of those images are likewise affected. Precisely what of the past *is* carried over from childhood to old age? That question is as significant for the oral historical geographer as the selective accession of documents is for the archivist.

Frank Zingel's Bega

Thus far, how the geographical memories of living witnesses might be tapped has been sought in a research literature that is rather disparate. It might focus the issues and round out the prospects to offer, from its inception, a vignette of just one exploration in oral historical geography. In 1963, while gathering material for a doctoral dissertation that probably shortened Spate's devotion to geography, I interviewed the late Frank R. Zingel of Bega, New South Wales, in an effort to compare his knowledge of country town land use with what the municipal rate books could document. When he died in 1977, he had spent all but two of his ninety-two years in Bega, an isolated rural service centre with barely 4,000 inhabitants, set amid coastal dairy farms, almost equidistant from Sydney and Melbourne. The vernacular cottage in which he was born was only 320 m from the Municipal Chambers in which he was to serve as mayor for seven terms, and alderman for altogether nineteen years. He was a carpenter and builder, who erected a home of his own next door to his birthplace, and spent his working life enlarging and observing the 'built environment' of his home town. Few of his contemporaries can have lived so spatially circumscribed a life in so simple a setting. Most of his thirteen siblings went walkabout. For my purposes, he was ideal.

From four interviews staggered over a period of three months, Frank Zingel produced fifteen foolscap pages of 'memory maps' comparing Bega in *c.*1911 with the Bega of 1963. In format, they consisted of page-long columns headed 'WAS' and 'IS', showing by matching line segments the land use for both sides of Bega's main commercial streets in *c.*1911 alongside what was visibly present in 1963. The *Bega Municipal Rate Book* for 1911 listed 681 separately rated land holdings; Zingel's maps identified 261 of them (38 per cent), at a time when the town had just kissed farewell to its Victorian pomp and circumstance, and when its population was 1,969. The street grid covered by these 'memory maps' had axes of approximately 900 m by 1,400 m, and accounted for 40 per cent (126 ha) of the town's built-over area. Zingel had no difficulty fitting his own outlines into the numbered series of portions, sections and lots used by the New South Wales Department of Lands to record and survey transactions in the property market. He was equally familiar with the details of easements, covenants and rights of way. Builders and aldermen know about such things. As an assembled mosaic, the fifteen maps covered the whole centre of the town and three detached, 'satellite' tracts where Chinese market gardeners had established themselves temporarily. The intervening suburban housing was omitted, as were peripheral farms on the flood-plain. Most of the maps took their orientation, not from the meridian of longitude, but from the corner of Gipps Street and Upper Street, where Zingel lived — at the centre of one man's Ptolemaic universe.

The *Bega Municipal Rate Book* for 1911 was undoubtedly but circumspectly accurate, and complete so far as it went. For each side of every street, it listed the land-owners serially, and alongside their names added the unimproved and improved values of the property, the names of any rate-paying lessee and such munificently informative descriptions as 'house and land'. Historical reconstructions of town land use from such particulars are attempted 'by guess and by God'. The penmanship, by contrast, was immaculate.

Although Zingel's maps did not always differentiate property owners from occupants, they were absolutely complete surveys of the area they covered, not missing so much as a hut, and their congruence with rate book entries was almost exact, plus or minus five years. Moreover, their glosses and marginalia brought the town to life. Except from oral testimony or other informal records, how does one ever discover that cricket was played in the horse paddock belonging to the Imperial Hotel? Such a fact belongs in the slighted category of 'unofficial' or 'secondary' land use. Where else does one learn that Walker Street was known as 'Irish Corner'? Rate books and newspaper advertisements do not say which downstairs workshops and stores had proprietors residing upstairs. Zingel's glosses repaired this defect, while also volunteering a number of building construction dates, the ages of some residents, the part-time occupations of many land-owners ('O. W. Mueller, photographer, and conductor of the Bega Orchestra'), and visual details of greater or lesser consequence and obscurity — such as the grandfather clock in the jeweller's doorway, or the monkey on its pole brought back from the Boer War.

Yet these seemingly trivial details, when pieced together, flood with light our understanding of how a small country town organised its affairs during that period, and how it felt to live there. They expose the names of Bega's 'establishment', the movers and shakers of their time. They make coherent the curious network of deliveries, commercial travellers and passing trade, of boardwalks over muddy footpaths and rights of way that reveal forgotten business connections. They define obsolete occupations, such as the town's one 'general secretary' whose office ran every club in Bega. They reveal what ignited or snuffed out major rearrangements of land use, such as the fires that destroyed Church Street stores and left Carp Street businesses supreme, or the unsettled estate of Daniel Gowing, which long delayed the redevelopment of prime real estate, or the 'local option' votes that revoked the liquor licences of half the town's hotels. We sense something of social attitudes when a Frank Zingel pointedly recalls three Chinese enclaves, and tells us that small boys knew some of them as Charlie Ah Kee, Andy Ting Lee, Bung-Eye and Old Joe. The official municipal rate books are more polite, and less veracious. Does any of this matter? It does, surely, to those interested in the historical geography of ordinary people.

Props

Historical geography written entirely from documentary evidence is like Shakespeare performed on an empty stage. Both need props. This is not to imply that all future research should abandon the archives and confine itself to oral evidence. What this advocate asks is only that historical geographers construct a little more of their stage scenery from oral testimony, and that they look more routinely for a *mutuality* of sources — written, printed, tabulated, heard or seen. As Spate liked to remind his postgraduate students, trying to find 'one key to open all locks' was the fondest futility. He also reprimanded us for structural weaknesses in what he called the 'architecture' of our theses: we were said to need props here or a whole scaffolding there. The image is a powerful one. Props hold together many a mutual assurance society, whether it be the glorious company of angels or the vainglorious company of scholars.

For the oral historical geographer, props can help to tease forth what is known about the geographical past. The tape-recorder need not be the only cue to that information. A faltering respondent might focus his thoughts and rekindle his memory if shown and asked to discuss an old map or annotate a new one. Old photographs or landscape paintings can be just as evocative of past associations, and equally fruitful in geographical research (Heathcote 1972). Nor should the stage setting be too stuffy, as the oral historians have learned from experience. Columbia University's first oral interviews took place in carpeted studios that looked like a vasectomy clinic. The participants were as formally attired, and probably as cordially communicative, as corpses during visitation. Nowadays, in the hollows of Appalachia, some old codger in a rocking chair will be chuckling away into a tape-recorder on his own front porch, being 'interviewed' by some teenage innocent, probably on his or her way home from the bowling alley. Yet both codger and corpse will be the 'key informant' for the project in hand — as relevant, as informed and as informing as the most crucial document in a collection.

One of these days, if any of this is believed, the program committee for a professional convention will organise the first session ever devoted to 'oral historical geography'. There, partisans will warmly dispute the advantages of interviewing individuals over group surveys, of tapes over transcripts. They will arrange nationwide schedules of collaborative research, arguing whether to distribute questionnaires (as the Kentucky Heritage Commission has) or to stop and question tourists at the Lincoln Memorial in Washington, D.C. (as did the Society for the North American Cultural Survey). Soon the fraternity will publish a journal, serving its own esoteric ends. Finally, inside a glass display case in some hallowed archive, there will be assembled a few scalps, and the earliest mementoes of those who pointed the way. I have no doubt that among them there will be a pair of suede desert boots, a bolo tie, the New Hebridean log of the S.S. *Rosinante,* and a plastic bust of Dr Endorn Weltsz.

Acknowledgments

I thank three Cincinnati friends and colleagues who supplied me with hints for this chapter — Ray Kennedy (Anthropology and Geography), Edgar Slotkin (Folklore), and Saul Benison (Oral and Medical History). For similar advice, I thank another friend and former colleague, Peter Topping of Harvard University's Dumbarton Oaks Center for Byzantine Studies. The chapter was read in pre-publication format before a gentlemen's drinking and moralising society, the Jolly Boys, whose 35 anonymous members I also want to thank for their unheeded criticism.

References

Baker, A.R.H. (ed.) 1972. *Progress in Historical Geography.* David and Charles, Newton Abbot.

Benison, S. 1965. Reflections on oral history. *American Archivist* 28: 71-7.

Bowen, E.G. and Carter, H. 1974. Preliminary observations on the distribution of the Welsh language at the 1971 census. *Geographical Journal* 140: 432-40.

Bowen, E.G., and Carter, H. 1975. The distribution of the Welsh Language in 1971: an analysis. *Geography* 60: 1-15.

Bynum, D.E. 1971. Oral evidence and the historian: problems and methods. *Journal of the Folklore Institute* 8: 82-4.

Cassidy, F. 1977. The dictionary of American regional English. (Paper read at Society for the North American Cultural Survey Annual Conference, October 1977, University of Louisville, Kentucky.)

Clark, C.M.H. 1973. *A History of Australia,* vol. 3. Melbourne University Press, Melbourne.

Dorson, R.M. 1964. Oral tradition and written history: the case for the United States. *Journal of the Folklore Institute* 1: 220-34.

Dorson, R.M. 1971. Introduction: folklore and traditional history. *Journal of the Folklore Institute* 8: 79-81.
Evans, G.E. 1971. *Tools of Their Trades — An Oral History of Men at Work c.1900.* Taplinger, New York.
Evans, G.E. 1976. *From Mouths of Men.* Faber and Faber, London.
Hart, R.A. and Moore, G.T. 1976. The development of spatial cognition. Pp. 258-81 in H.M. Proshansky, W.H. Ittelson and L.G. Rivlin (eds.), *Environmental Psychology — People and Their Physical Settings.* Holt, Rinehart and Winston, New York.
Harwood, F. 1976. Myth, memory, and the oral tradition: Cicero in the Trobriands. *American Anthropologist* 78: 783-96.
Heathcote, R.L. 1972. The artist as geographer: landscape painting as a source for geographical research. *Proceedings Royal Geographical Society of Australasia, South Australian Branch* 73: 1-21.
Ittelson, W.H. (ed.) 1973. *Environment and Cognition.* Seminar Press, New York.
Jack, R.I. 1977. Local history in Australia. *Current Affairs Bulletin* 54(2): 24-30.
Law, R.G. 1969. Pits and kumara agriculture in the South Island. *Journal of the Polynesian Society* 78: 223-51.
Lewthwaite, G.R. 1970. The puzzle of Tupaia's map. *New Zealand Geographer* 26: 1-19.
Lowenthal, D. 1975. Past time, present place: landscape and memory. *Geographical Review* 65: 1-36.
Lowenthal, D. 1977. The bicentennial landscape: a mirror held up to the past. *Geographical Review* 67: 253-67.
Lowie, R.H. 1915. Oral tradition and history. *American Anthropologist* 17: 597-9.
Lowie, R.H. 1917. Oral tradition and history. *Journal of American Folklore* 30: 161-7.
McCourt, D. 1972. The use of oral tradition in Irish historical geography. *Irish Geography* 6: 394-410.
Meckler, A. and McMullin, R. (eds.) 1975. *Oral History Collections.* Bowker, New York.
Newcomb, R.M. 1972. The nostalgia index of historical landscapes in Denmark. Pp. 441-3 in W.P. Adams and F.M. Helleiner (eds.), *International Geography,* vol. 1. University of Toronto Press, Toronto.
Pendergast, D.M. and Meighan, C.W. 1959. Folk traditions as historical fact: a Paiute example. *Journal of American Folklore* 72: 128-33.
Pocock, D.C.D. 1976. Some characteristics of mental maps: an empirical study. *Transactions of the Institute of British Geographers* (New Series) 1: 493-512.
Powell, J.M. 1972. The records of the New Zealand Lands and Survey Department. *New Zealand Geographer* 28: 72-7.
Roder, W. 1977. An alternative interpretation of men and women in geography. *Professional Geographer* 29: 397-400.
Sack, R.D. 1978. Geographic and other views of space. Pp. 166-84 in K.W. Butzer (ed.), *Dimensions of Human Geography.* University of Chicago, Department of Geography, Research Paper 186.
Schlesinger, A., Jr 1971. The historian as participant. *Daedalus* 100: 339-58.
Shackelford, L. and Weinberg, B. (eds.) 1977. *Our Appalachia — An Oral History.* Hill and Wang, New York.
Spate, O.H.K. 1954. *India and Pakistan — A General and Regional Geography.* Methuen, London.
Spate, O.H.K. 1956. Bush and city: some reflections on the Australian cultural landscape. *Australian Journal of Science* 18: 177-84.

Spate, O.H.K. 1957. Manuel Godinho de Erédia: quest for Australia. *Meanjin Quarterly* 16: 109-22.

Spate, O.H.K. 1961. On being an expert. *Quadrant* 5(18): 25-34.

Spate, O.H.K. 1965. Progress at Mbananakoro. Pp. 66-87 in O.H.K. Spate, *Let Me Enjoy: Essays, Partly Geographical.* Australian National University Press, Canberra.

Starr, L.M. 1971. Oral history: problems and prospects. Pp. 275-304 in M.J. Voigt (ed.), *Advances in Librarianship,* vol. 2. Academic Press, New York.

Storm-Clark, C. 1978. Review of P. Thompson. The Voice of the Past: Oral History. *Listener* 100: 190-1.

Strehlow, T.G. 1936. Notes on native evidence and its value. *Oceania* 6: 323-35.

Vansina, J. 1965. *Oral Tradition: A Study in Historical Methodology.* Aldine, Chicago.

Vansina, J. 1971. Once upon a time: oral traditions as history in Africa. *Daedalus* 100: 442-68.

Walker, R.B. 1964. Bushranging in fact and legend. *Historical Studies Australia and New Zealand* 11: 206-21.

Chapter 6

City of Magnificent Distances

*Larry Sternstein**

> Bangkok . . . is a city of magnificent distances, especially if the tide happens to be against you both going and returning. (Cort 1886.)

Rarely may a would-be restorer of a chartless cityscape do more than plait chance traces from keen contemporary, perforce alien eyes. The following word-picture of centennial Bangkok has been pieced together from the laundered leavings — descriptions freed from wilful favour and wild fancy — of American and European residents of the city during the latter half of the nineteenth century.

> Bangkok is built for about six miles on both sides of the noble river Chow Paya [Chao Phraya] . . . thirty miles from the mouth . . . The downward current . . . is very strong, but the rising tides force water back into the creeks and canals that intersect . . . all over this semi-aquatic city . . . which depends more on her water-ways . . . than on streets of solid ground.
>
> From the centre of the broad deep stream stand out the tall masts of large . . . steamers, their huge hulls towering far above the . . . innumerable smaller . . . native craft . . . of every

* *Larry Sternstein is Senior Lecturer in the Department of Geography, School of General Studies, The Australian National University. After graduating from City University, New York, and Syracuse University, New York State, he took his doctoral degree in the Department of Geography, Research School of Pacific Studies, The Australian National University, in 1965.*

variety . . . that ply between them and the shore . . . Chinese junks . . . moored by immense ratan cables . . . also scores of steam-yachts . . . The river . . . is by no means a 'silent highway' . . . dozens of small skiffs are flitting about, 'manned' by women . . . bargaining . . . and disposing of their fruits and vegetables, their firewood and varied up-country produce . . . rice, sugar, salt, cotton, oil, dried fish, or dye-wood . . . Here and there are . . . little boats where John Chinaman sells . . . a frugal meal of curried rice, boiled vegetables, bits of pork, or dried fish and cakes . . . There are . . . private pleasure boats . . . Boating is . . . the only chance of getting a little fresh air, and nearly every one keeps a boat.

Along the shore . . . in rows five or six deep . . . the native boats, with their deck-house covered in with a semicircular roof, under which the boatman and his family make their permanent home . . . Thousands of the people live in floating houses . . . lining both banks of the river . . . They are but one storey high . . . built of . . . light wood . . . thatched with the leaves of the attap palm . . . and placed in rafts of large bamboos, which rise and fall with the tide . . . moored to the bank, or to . . . large posts on each side driven deep into the muddy bed of the river . . . Many of them are open in front with a veranda, and are shops . . . Here you can purchase your supplies . . . by stopping your boat and pointing out what you want in the wide open room before you.

Beyond these, on either bank . . . stretches a wide expanse of the sloping roofs of the native houses . . . relieved at short intervals by the glittering spires of temples . . . or the pinnacles of the royal palaces, rearing their heads high above everything . . . and on a sunny day the effect of the glistening towers, many of them gilt to the very top, is very beautiful.

Really imposing as the view . . . from the river is, the contrast on landing . . . is very great.

Many roads have been constructed during the last few years, it is true . . . but they are all below the flood-level . . . the city being built on the edge of a great alluvial plain, which is inundated during the rainy season, so that the roads will have to be raised to be of any permanent advantage . . . The main road . . . which is several miles in length, is itself often partly under water during the south-west monsoon . . . The streets are nearly all very narrow and crooked . . . They are also very uneven, with here and there great holes, crossed by rough stepping stones, and during the highest tides are overflowed with water, sometimes knee-deep, for several hours each day . . . The back lanes and bypaths . . . are in a chronic state of filth, wet or dry . . . In the city proper . . . which is walled in . . . are, however, a few wide pleasant streets . . . long enough to furnish six or seven miles of good driving. They are kept in tolerable repair . . . daily swept . . . in the dry season sprinkled . . . to lay the dust, and at night lighted by lamps and patrolled by watchmen.

The facilities for locomotion are . . . insufficient . . . within the last year or two hacks have been running . . . The carriages, kept for hire by a few Klings [Dravidians] and Malays, are generally in a most dilapidated state, while the horses are still worse.

The nobles . . . have erected a great many handsome brick houses, which are planned by European architects, and are roomy and comfortable . . . and some of them elegantly furnished with English, French and Chinese furniture . . . The missionaries, foreign consuls, merchants and wealthy Chinese have good, substantial dwellings . . . on the banks of the river . . . to avoid the not too savoury smells of the interior of the city . . . and house-rents there are accordingly high . . . The middle class dwell in houses built of wood, usually unpainted teak, and roofed with earthen tiles. They are small and illy ventilated, and here the people huddle together, from the parents to the children of the third and fourth generation . . . They have very little furniture . . . The lower class live in huts made of woven bamboo, and thatched with leaves . . . Nearly all dwellings are built on posts . . . which elevate them five or six feet from the ground, and are reached by ladders, which at night are often drawn up to prevent dogs or thieves coming into the house.

The natives consider it an insult if they enter another's house and are not invited to eat betel, and it is equally impolite to refuse the proffered cud. Indeed, it occupies so important a place in the economy of their social life that a wedding is called . . . literally 'betel-tray' . . . Betel-chewing disfigures the mouth wonderfully, causing the teeth to protrude and blacken, and the lips and tongue to crack . . . The cud they chew so persistently is a combination of ereca-nut, cera-leaf, lime, tobacco, camphor, and tumeric. It is . . . indulged in by all classes . . . It costs almost as much as their food, especially among the poor.

Housekeeping and needlework form so small a part of female labor here that much opportunity is given for out-of-door work . . . Women enjoy greater liberty than in almost any other Oriental land. You meet them everywhere; and in the bazaars and markets nearly all the buying and selling is done by them . . . They are seen performing all sorts of labor.

Of the city's half million people, perhaps nearly one-half are Chinese, Hindoos, Malays, and other foreigners, of whom less than three hundred are Europeans . . . The Chinese have been in Siam since time immemorial, and have increased . . . There is no census taken . . . and even the government has no positive means of knowing the number of inhabitants. But . . . every steamer and sailing craft from China is swarming with . . . Celestials.

The Chinese are very clannish, and settling together have given to some portions of the city quite a 'Celestial' appearance. Their streets are close and dark, some of them covered overhead, and filled with real Chinese odors . . . Pigs, dogs, cats and children throng these thoroughfares; there is no danger from horses or carriages, the streets are far too narrow to admit them . . . The houses in the markets are so made that the front can all be taken down in the daytime, and the whole inner room and its contents exposed . . . If they have counters, they fill nearly the whole room, which is often not more than ten by eight feet, and . . . the merchant sits in the midst of his goods. The whole family . . . lives in or back of the store . . . All sorts of trades are carried on in sight of every passer-by . . . At a restaurant or bakery . . . you see the inmates prepare the food or cake before your very eyes. The blacksmiths and tinners are hammering for dear life; and the tailors cutting and sewing, but always on the alert to sell a needle or a half dozen buttons . . . On all sides are Chinese joss-houses, Chinese carpenters' shops, Chinese cabinet works, Chinese carriage manufacturers; wherever there is any work going on it is sure to be under the sign of a Chinese proprietor, though here and there may be seen a Siamese pottery works, where the brittle portable ovens, pots, and water-bottles are produced . . . In happy confusion may be seen Chinese pawnbrokers' shops; Siamese and Chinese eating-houses . . . street stalls . . . Chinese duck farms, where the ducklings are reared from the egg, and find plenty of rubbish to rout about in . . . Chinese drug-shops . . . dram-shops, where . . . the liquor-jars . . . are out in the street, and the people stop and drink . . . Native arrack . . . made of molasses . . . costs but little . . . The home production is not equal to the demand, and . . . liquor is imported from China, Batavia, Singapore, and Europe . . . Eating and smoking opium are also on the increase and the law which threatens all consumers of it with confiscation of property and death, is not now enforced. There is a weed . . . (which I think is Indian hemp), grown abundantly in Siam, and those who are too poor to buy opium use this instead . . . It would be hard to find a Siamese who did not use tobacco in some form . . . All chew the weed with their betel, and some use it as snuff . . . More numerous perhaps, than anything else, are the ubiquitous gambling-houses . . . probably many hundred in Bangkok . . . with a Chinese . . . theatre, close at hand . . . The gambling establishments are all in the hands of the Chinese . . . Gambling, like many other things in Siam, is a monopoly, and the government sells to the highest bidder the privilege of licensing and controlling all such establishments . . . They afford no small amount of revenue.

Next to gambling-halls, theatres, and perhaps dram-shops, the structures which are most numerous . . . are the temples . . . Bangkok may be termed the City of Temples . . . There are between one and two hundred temples in the city . . . No one can be long in Siam without being astonished at the large part which the wat [temple] occupies as a social centre in the every-day life of the people . . . The kings and nobles . . . and the . . . people spend vast sums . . . on these temples . . . They occupy the pleasantest parts of the city . . . A Siamese wat . . . consists of a number of buildings scattered about a large park-like enclosure . . . Pagodas, or sacred spires — detached pyramidal piles of solid masonry, frequently reaching a great height — are always found in connection with the temples. These . . . contain some relic of Buddha, and are sacred to his memory . . . Each wat has . . . its . . . preaching hall. Each wat has also its library, containing the sacred books of Buddhist scriptures . . . But a . . . wat is not merely a place of worship; most of all it is a . . . monastery . . . Priests are easily recognised by their yellow robes and shaven heads . . . It is said that there are ten thousand . . . in Bangkok . . . and one can easily believe it for you meet them everywhere.

When Bangkok was founded in 1782 the king's palaces and all the principal buildings were erected on the left [sic] bank of the river, but now the royal palaces . . . are on the other side . . . within the walls of the city . . . Within the city walls are palace walls . . . about a mile in circumference . . . thick and high, with double doors and . . . forts, and in this most carefully guarded enclosure are the grand royal palaces of the king and queen . . . the audience halls, the mint, arsenal, halls of justice, museum, royal chapel, and separated from them by an inner wall is the royal harem . . . that 'City of Women' . . . which is in itself a compact little town, with several streets, a bazaar, a temple, pleasure-gardens and the homes of the numerous wives, sisters and other relatives of the king . . . His majesty . . . has just completed the new royal residence . . . The style is a mixture of different schools of European architecture, the picturesque and characteristic Siamese roof, however, being retained. The internal fittings . . . are most elaborate . . . the most costly furniture having been imported from London . . . and Paris . . . and illuminated with electric lights.

The work which, in popular estimation at least, will make his Majesty's reign most memorable . . . is the completion . . . of the great royal temple . . . for the Emerald Buddha, the palladium of the capital . . . On the 21st. of April 1882, the ceremony of final dedication . . . of this magnificent pile of buildings . . . was performed . . . to give the city its crowning glory . . . on the hundredth anniversary of the capital of Siam . . . the celebration of the centenary of the city was marked by the inauguration of many reforms, many of which are personified in the public buildings then commenced or already erected for their administration . . . Bangkok has few sanitary laws, and such as have been enacted are seldom strictly enforced; yet . . . the sanitary condition of Bangkok is . . . receiving great attention, and measures are being gradually adopted to mitigate the severity of the fearful epidemics of cholera, which in times past have . . . devastated the country.

Not one half century ago Siam was sealed against the entrance of . . . the dreaded . . . Europeans . . . Chinese junks . . . laden with stones . . . were sunk at the mouth of the river, and chains stretched from shore to shore to prevent the . . . 'fire-ships' . . . steaming up to Bangkok . . . To-day she is in treaty relations with all . . . countries, and . . . now, large steamers at high tide cross the bar and sail on and on through a wide, deep, open river and cast anchor in the very heart of the . . . capital . . . In the city . . . there are large business-houses conducted by foreign merchants . . . Good inducements are offered to foreigners to enter the various departments of trade, and full protection is given.

The Chinese . . . fare better in every respect than the ordinary natives . . . Their privileges are greater, and their taxes less oppressive . . . They are not drafted into the army, and are not subject to the beck and call of every officer and nobleman in the unlimited ranks above them . . . Chinese of wealth often become favorites with the rulers and receive titles of nobility, and these noblemen in return present their daughters to their majesties . . . Although a Chinaman may have left a wife in his native land, that does not prevent him taking as many as he can support . . . A Chinese woman is rarely seen in Siam.

The Chinese element in Siam is a powerful one. No other race can compete with it, not even excepting the Caucasian . . . They have their temples and joss-houses, their religious fetes, processions, and festivals. Their holidays are recognised by the Government . . . Their secret societies are many and formidable . . . the natives of each province [of China] holding together and working to promote the interests of their own particular clan. They have frequent quarrels . . . which sometimes threaten the peace of the kingdom and often disturb that of the city . . . We find the Chinese in every business. They are . . . industrious and enterprising, and do most of the work and control much of the wealth of Bangkok, outside of what is in the hands of the kings and nobles . . . In the days when Siam had a sailing fleet of merchantmen the owners were principally Chinese, as were also the shippers and crews . . . Since steamships have been introduced we find that the owners and agents of some of these are Chinamen. The saw-mills and rice-mills worked by muscle power are all owned by Chinese, and since the introduction of steam-mills they are not slow to adopt these modern improvements, so that now several steam saw-mills and rice-mills are owned by enterprising Chinamen . . . They are . . . gardeners, shopkeepers, carpenters, bricklayers, tailors, sailors, shoemakers, blacksmiths, fishermen and washermen . . . All the mills employ Chinese coolies; all cargo-boats for loading and unloading ships are manned by these coolies.

As gardeners the Chinese are very successful . . . Bangkok is situated upon a plain which is almost at sea level; there are no hills or mountains to relieve the eye in any direction. This plain is covered with rice fields, vegetable gardens, fruit and spice orchards, and gardens of betel, cocoanut, banana, and other tropical trees; orange, mango, coffee, and nutmeg . . . The land is made sufficiently dry by throwing it up in large beds ten to twelve feet high . . . The deep ditches between have a supply of water even in the dry season . . . The gardener lives within the premises, his . . . hut guarded by a multitude of dogs and a . . . stench of pigsty . . . The Chinese do not cultivate the paddy fields to any great extent, but buy the rice from the producers and bring it to the Bangkok markets. The seri-leaf . . . is cultivated with great care . . . in the betel-gardens . . . This leaf . . . bright green, tender and juicy . . . covered with a pink lime paste and a little tobacco and betel-nut added . . . rolled up cross-wise and chewed . . . is used . . . extensively in Siam . . . Rotten fish is used as a fertiliser, and consequently the breezes which blow over these gardens are not 'spicy breezes', but . . . very offensive, obliging one in passing to suspend respiration for a time . . . But if you are so fortunate as to be gliding through . . . canals in other parts of the city, where the flower-gardens and orange trees are blooming in beauty and fragrance, every sense is filled with delight, and one could almost wish they might drift through such enchanting ways forever.

(Anonymous 1884a, 1884b; Bacon 1892; Bock 1884; Bradley 1884; Caddy 1889; Cort 1884, 1886; House 1884; Mrs House 1884; McDonald 1884; Vincent 1884.)

Is this anecdotal account accurate?

To censure diplomats, merchants and evangelists for leaving vivid impression instead of balanced assessment would be hypercritical. Nevertheless there are several questions which can rightly be raised about reportage of this kind. One such query concerns the wherefore of several of the traits delineated. For example, betel-chewing among the Siamese is condemned as a disgusting habit or an expensive luxury, and scorned; but no one seems to have troubled to discover why betel-chewing had been ritualised. In fact, the active constituents of the betel-cud have antiseptic, carminative, cathartic and tonic properties (Fluckiger and Hanbury 1879). Small wonder, then, that extensive, carefully cultivated betel gardens surrounded Bangkok, there was brisk trade in the constituents of the betel-cud, and each household had a betel-service comprising a box, a tray and a spittoon, to accept the outpour from the salivary glands which are activated wonderfully when chewing. Would our informants' report of the pervasive practice of betel-chewing have been different had they known of its benefits?

Another question which may be asked is whether the word-picture, though valuable, is accented peculiarly and portrays centennial Bangkok in caricature. Caricature can be incisive. Whether the anecdotal anthology is incisive or merely grotesque might be assessed, in the first instance, by comparison with the choice reminiscences of keen observers at other times. Eyewitness accounts of Bangkok a generation either side of its centennial have been compiled (Sternstein 1976), spanning the early nineteenth to the early twentieth centuries. Taken together, these suggest there is nothing freakish in the description of centennial Bangkok. However, one particular incongruity does show up.

The population of Bangkok in the 1880s commonly was estimated at a half-million, though some thought a million nearer the mark. A generation earlier, in the 1850s, the population of Bangkok was said to approximate 350,000. During the opening decade of the twentieth century, the population of Bangkok was thought to approximate 500,000, though some reported the number at not more than 400,000 and others as not less than 600,000. A half-million hearts in centennial Bangkok may be judged excessive, since the population of the capital had burgeoned after the late 1880s. How excessive is the estimate cannot be gauged from the reportage. A centennial count is needed.

On 21 September 1883, His Majesty, King Chulalongkorn, expressed surprise over the use of the newly established postal service which had exceeded expectations *(Siam Weekly Advertiser* 22 September 1883). Collection boxes had been set up throughout the city a few months before. Their bottoms were reportedly made of

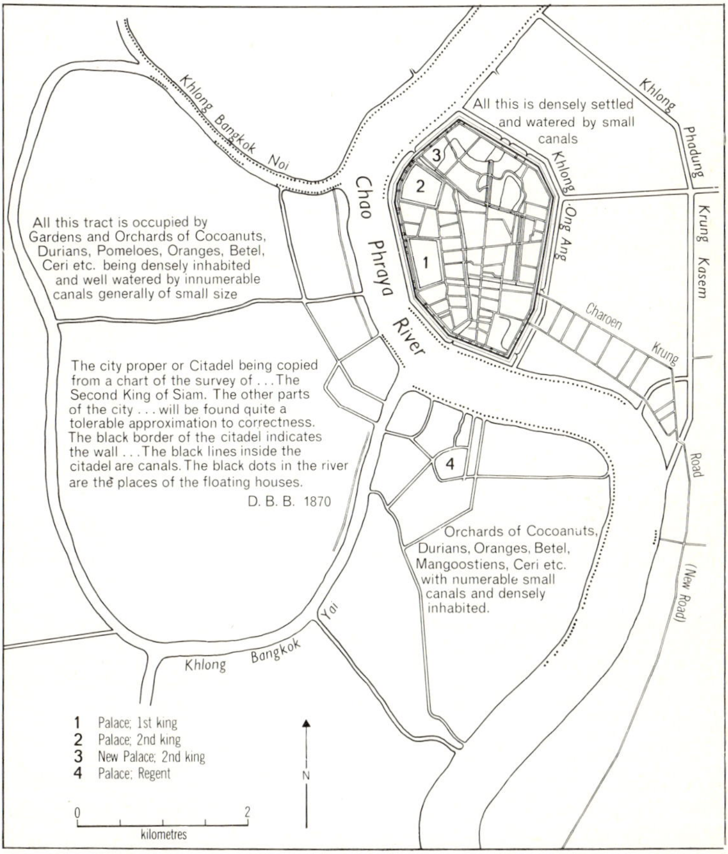

Figure 6.1 The lineaments of the city *c.* 1870 from the first unstylised but annotated map of Bangkok by D. B. Bradley (Mission Press, Singapore).

sandal-wood 'to impart fragrance to the missives, and thereby cultivate a taste for letters' (Cort 1886), though the anonymity of the post triggered an avalanche of abuse addressed to certain nobles, and also the forwarding of heavy packages of trash for which delivery charges were due. To facilitate the mails, each house was given a number, surnames were assumed by each family, and a roll of names and addresses was compiled in 1882. I discovered this roll in the National Library of Thailand. It contained information about each dwelling and household which exceeded my expectations and provided a substantial basis from which to chart the cityscape at that time. Now only several months into the quinquennium needed to sort it out, I can but saucer up some cream.

According to the postal roll, which was more a census than a roll in that it included the number of persons in each household, the population of Krung Thep — comprising the city of Bangkok and a wide area surrounding, roughly thrice the area shown on Figure 6.1 — approximated 169,000 in 1882 (Table 6.1). This sum seems reasonable in relation to subsequent censuses, though it differs

Table 6.1. Population of Krung Thep and the city of Bangkok 1882*

Nationality Group	Krung Thep	City of Bangkok
Thai	136,000	93,000
Chinese	27,000	23,000
Malay	4,000	1,800
Indian	1,000	700
Westerner	300	300
Other	1,000	900
Total	169,300	119,700

* Figures are rounded because only estimates could be made for several sub-populations (for example, monks and harem dwellers) omitted from the postal roll.

markedly from mid-nineteenth century estimates and contradicts previous analyses (Skinner 1962; Sternstein 1966). It permits subsequent, unprecedented population growth stemming from an inrush of immigrants (mainly Chinese) and in-migrants, as well as a downrushing mortality rate which came with the introduction of preventive medicine and the improvement of sanitary conditions in the capital. The incontrovertible argument in support of this new thesis is shown graphically on Figure 6.2 and outlined in its caption.

How could long-term residents have arrived at the gross overestimate of a half-million for the population of centennial Bangkok? Perhaps our eyewitnesses succumbed to a visual fallacy. As they meandered through the city and its environs along innumerable waterways, and the few roadways, they saw houses closely built either

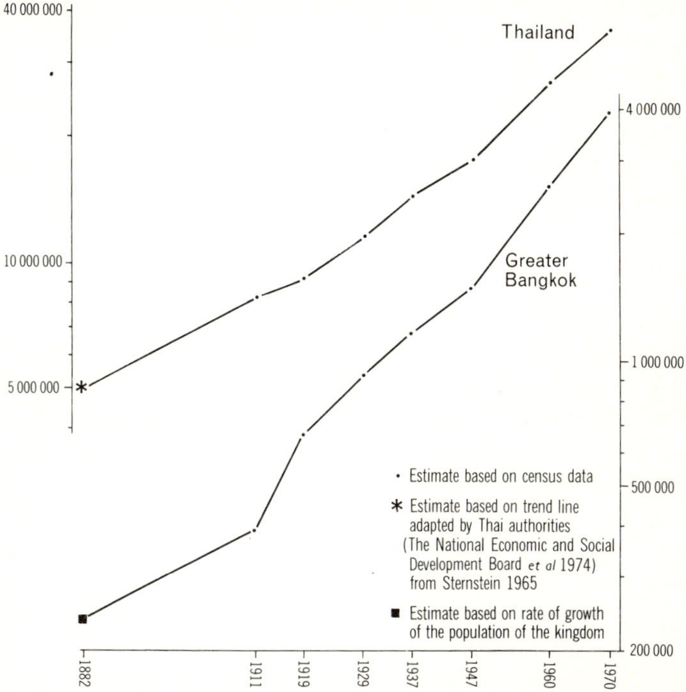

Figure 6.2 Growth of the population of Thailand and Greater Bangkok 1882 to 1970. Greater Bangkok comprises provinces Krung Thep Maha Nakhon, Nonthaburi and Samut Prakan — an area roughly forty times that shown on Figure 6.1 but the smallest for which data are available for the period 1882 to 1970 (Sternstein 1964) and within which the area canvassed in the centennial count is wholly included. In 1882 Greater Bangkok held a population of 250,000 provided numbers had increased from 1882 to 1911 at the rate of growth for the kingdom as a whole. If the rate of growth in Greater Bangkok were greater than that for the kingdom — certainly it has been so since — the population of Greater Bangkok in 1882 was less than 250,000. We need not guess how much less; 250,000 people in an area more than ten times that included in the postal census must accredit the centennial return of 169,000 for the city and its environs.

side and possibly attributed similarly dense housing to land between the ways; in fact these tracts were not built-up. Perhaps, also, the great number of native trading boats in the city, in which lived the boatmen and their families, may have been counted as permanent city residences by our informants, though few were so certified by the postal authorities. Possibly, too, the estimated population of Bangkok in the 1850s of 350,000, having been made by several sensible men, induced our informants to propose a 'reasonable' figure of a half-million souls in the city a generation later.

Foreign residents of centennial Bangkok believed that non-Thai made up half the population of the city; the postal roll puts the non-Thai fraction of Krung Thep at a fifth (Table 6.1). Although eyewitnesses probably saw relatively more non-Thai than there were (undue weight being given particularly to the pigtail-wearing Chinese, who stood out boldly from the crowd, engaged in eye-catching activities, and massed near the consulates and business houses of the Europeans), the great difference between the non-Thai fractions of the population reported may reflect different definitions of the city. Beholders of centennial Bangkok limited the city vaguely to several miles along both banks of the Chao Phraya centred on the citadel. The stretch of river shown on Figure 6.1 appears to be a fair representation of their description, though overmuch space may be shown west of the Chao Phraya. The city was uncommonly difficult to bound because of its squiggly and discontinuous configuration which matched its framework of waterways. Also confounding, perhaps, was the juxtaposition and the intermingling of agricultural and non-agricultural households; indeed, within households individuals engaged in farm and off-farm activities. If the centennial city of Bangkok covered two-thirds of the area shown on Figure 6.1 centred on the Chao Phraya river, its population, according to the postal roll, totalled 120,000 approximately a quarter of which was non-Thai (Table 6.1). Annotations on Bradley's 1870 map of Bangkok (see Fig. 6.1) describe sizable tracts as densely inhabited orchards and gardens. According to the postal roll fully a fifth of household heads in the 'city' worked the land. Of those who did not work the land better than two-fifths were non-Thai, which fraction is not too far from the anecdotal nearly half non-Thai in the population, albeit the visible population only, of centennial Bangkok.

Early in the nineteenth century, eyewitnesses put the non-Thai fraction of the population of Bangkok at two-thirds and the number of Chinese at greater than the Thai. Skinner (1962) has argued that this seemingly odd situation arose since the Thai were not free to move to their capital city because they were anchored to the up-country estates of their patrons and masters, whereas the Chinese immigrants, whose main point of entry was Bangkok, remained outside the semi-feudal system. In truth, however, the obligation of client to patron had been progressively eased and monetised since the reign of Rama II (1809-29) and, more importantly, clients had side-stepped formal, local obligations always. Moreover, only male clients from the age of twenty were required for corvée duties. So it is not surprising that the centennial count found more Thai than Chinese in Bangkok in 1882, despite the fact that during the latter half of the century, state control of trading was whittled away which, together with the conclusion of treaties with many European nations, boosted the economy and increased immigration, particularly of the Chinese.

The reportage about centennial Bangkok makes little reference to the distribution of the population other than that certain sections had a 'celestial appearance'. From the postal roll it is clear that the

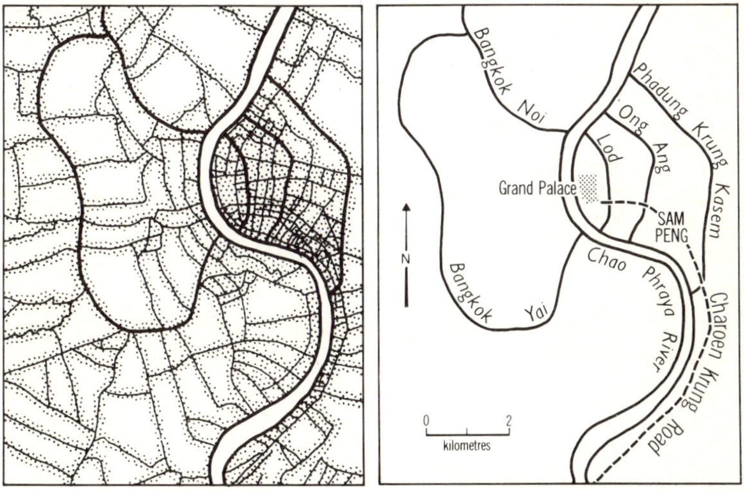

Figure 6.3 Distribution of the population of Bangkok 1882. This map was made by representing the main lines of settlement along waterways and roadways and by photographic reduction running these together to give an impression of population density from the relative darkness of the resulting patterns.

population of centennial Bangkok was aligned strictly along reticulated waterways and the few roadways in a meshwork; here tightly drawn, there loosely made, Thai were everywhere. Chinese were as widely distributed as the Thai though less uniformly; their distribution, in fact, is described rather nicely by the pattern of population density shown on Figure 6.3. Two-thirds of the Chinese resided immediately east of the Chao Phraya river; little less than a third in Sam Peng (if this quarter is delimited generously as the whole of the area along the river between the southern stretches of the Ong Ang and Phadung Krung Kasem canals); nearly a fifth were within the walls of the citadel, a goodly number being adjacent to the walls of the Grand Palace itself; a tenth lined the river south of Sam Peng; and others fronted the waterways and 'roadways' north of the citadel. The Chinese west of the Chao Phraya favoured major waterways, particularly the river bank.

The concentration of Chinese in Sam Peng may have rather a subtle explanation. When Ayutthaya, the former capital of the kingdom, was abandoned and the new capital was set up on the west bank of the Chao Phraya in 1767 opposite the citadel-to-be on the east bank, Bangkok may have been more a Chinese than a Thai settlement. The site of the as yet unmade citadel was occupied by a rich Chinese merchant who, together with a considerable community of Chinese, was invited to move to an uninhabited area immediately beyond the walls of the new citadel (Wenk 1968). This choice commercial river frontage had been uninhabited (actually under fruit trees) because it lay within the field of fire of fortress Bangkok (depicted on a map in Volant des Verquains 1691) which had guarded the riverine approach to Ayutthaya approximately a hundred kilometres upstream. When Bangkok succeeded Ayutthaya as capital, its riverine approach was guarded by forts near the mouth of the Chao Phraya.

Other national groups of size in centennial Bangkok — Malays, Indians and Westerners — were in nests in the more accessible parts of the city. The largest concentration was along the river immediately south of Sam Peng; half the Westerners and more than a third of the Malays and the Indians were here. A second nucleus in the heart of Sam Peng held a quarter of the Westerners and the Indians but relatively few Malays. A third next lined the west bank of the river opposite and south of the Grand Palace; here were a third of the

Table 6.2. Occupation of head of household by nationality, city of Bangkok 1882 (per cent)

Occupation*	Thai	Chinese	Malay	Indian	West-erner	Other
Farmers and fishermen	25	12	13	2	—	31
Makers of products	13	16	23	6	2	7
Labourers and construction workers	6	9	9	..	—	4
Transport and communications workers	1	..	5	6	17	3
Clerical workers	8	3	4	3	8	1
Sales workers	14	40	22	66	10	9
Service workers	7	6	6	6	3	4
Professional and technical workers	5	2	3	3	27	2
Administrators and managers	5	3	3	1	28	2
Armed forces	7	..	2	—	—	33
Other and unknown	9	9	10	7	5	4
Total	100	100	100	100	100	100

* Occupation refers to the kind of work performed, not the industry in which a person worked.
.. Less than 0.5 per cent.

Indians and a quarter of the Malays but few Westerners. Pure ethnic stands were few and small; two such stands each comprised a troop of Vietnamese mercenaries charged with the operation of heavy artillery. The gathering together of members of different national groups reflected a common concern with trade.

Although trade was a major activity for each of the several major ethnic groups in centennial Bangkok, the membership of each group engaged in a wide range of occupations. There was a certain occupational specialisation within national groups, but no occupational exclusiveness (Table 6.2), despite contemporary assertions to the contrary (Bock 1884; Cort 1886; Newman 1883). There was not a product made nor an activity pursued which was entirely in the hands of a single national group.

So Furnivall (1939) was unwise to extend his famous thesis concerning the development of a 'plural society' resulting from the impact of colonialism and capitalism on certain tropical polities by pointing to 'a plural society also in independent states, such as Siam, where Natives, Chinese and Europeans have distinct economic functions, and live apart as separate social orders'.

Figure 6.4 River-level view of Bangkok downstream from the northern end of the Grand Palace. The tower on the west bank is the *phra prang* (pagoda) of Wat Arun. (*Source:* Anonymous 1884b.)

Doubtless, Furnivall's distance from Bangkok lent enchantment to his view. Even residents of centennial Bangkok found it difficult to find an appropriate perspective. For example, the *phra prang* (pagoda) of Wat Chang, now named Wat Arun (Temple of Dawn), was regarded as rather a tawdry construction because of the crude materials used in its ornamentation, including bits of broken pottery and crockery (Fig. 6.4). This was Bock's assessment (1884), who valued the *phra prang* because it afforded him a bird's eye view of the city. The tower was not built for that purpose, but was to be viewed across the Chao Phraya river from the Grand Palace, specifically when the spire with its reflective bric-à-brac caught the rays of the rising sun. Fashioned from materials chosen carefully and arranged deliberately, the *phra prang* of Wat Arun is a great work because it is set at a magnificent distance.

Marc Bloch (1954) reminds us that 'each science has its appropriate aesthetics of language'; that 'there is no less beauty in a precise equation than in a felicitous phrase'. Each limb of learning has, also, its appropriate aesthetics of perception, a magnificent distance from which objects under study are in focus. I find that geographers commonly study objects at big, not magnificent, distances. I believe, however, that once we are agreed on the aesthetics appropriate to our craft we will build a city of magnificent distances.

References

Anonymous 1884a. The Indo-Chinese Peninsula. Pp. 15-80 in *Siam and Laos, as Seen by Our American Missionaries*. Presbyterian Board of Publication, Philadelphia.

Anonymous 1884b. The wats of Siam. Pp. 269-303 in *Siam and Laos, as Seen by Our American Missionaries*. Presbyterian Board of Publication, Philadelphia.

Bacon, G.B. 1892. *Siam, the Land of the White Elephant, As It Was and Is*. Scribner, Armstrong, New York.

Bloch, M. 1954. *The Historian's Craft*. University Press, Manchester.

Bock, C. 1884. *Temples and Elephants: The Narrative of a Journey of Exploration through Upper Siam and Laos*. Sampson, Low, Marston, Searle and Rivington, London.

Bradley, D.B. 1871. *The Bangkok Calendar*. American Missionary Association, Bangkok.

Bradley, D. 1884. A gambling establishment. Pp. 233-5 in *Siam and Laos, as Seen by Our American Missionaries*. Presbyterian Board of Publication, Philadelphia.

Caddy, F. 1889. *To Siam and Malaya in the Duke of Sutherland's Yacht 'Sans Peur'*. Hurst and Blackett, London.

Cort, M.L. 1884. Housekeeping in Siam. Pp. 175-83 in *Siam and Laos, as Seen by Our American Missionaries*. Presbyterian Board of Publication, Philadelphia.

Cort, M.L. 1886. *Siam: or, the Heart of Farther India*. Randolph, New York.

Fluckiger, F.A. and Hanbury, D. 1879. *Pharmacographia, A History of the Principal Drugs of Vegetable Origin, Met With in Great Britain and British India* (Second edition). Macmillan, London.

Furnivall, J.S. 1939. *Netherlands India, A Study of Plural Economy*. Cambridge University Press, Cambridge.

House, S.R. 1884. Cholera-times in Bangkok. Pp. 241-6 in *Siam and Laos, as Seen by Our American Missionaries*. Presbyterian Board of Publication, Philadelphia.

House, S.R. Mrs 1884. Sight-seeing in Bangkok. Pp. 81-95 in *Siam and Laos, as Seen by Our American Missionaries*. Presbyterian Board of Publication, Philadelphia.

McDonald, N.A. 1884. The Chinese in Siam. Pp. 146-61 in *Siam and Laos, as Seen by Our American Missionaries*. Presbyterian Board of Publication, Philadelphia.

Newman, W.H. 1883. Siam Consular Report for 1882. *Annual Diplomatic and Consular Reports on the Trade of Siam*. Great Britain, Foreign Office, London.

Skinner, G.W. 1962. *Chinese Society in Thailand: An Analytical History*. Cornell University Press, Ithaca.

Sternstein, L. 1964. Settlement in Thailand, patterns of development. Unpublished Ph.D. thesis, The Australian National University, Canberra.

Sternstein, L. 1966. The distribution of Thai centres at mid-nineteenth century. *Journal of Southeast Asian History* 7: 66-72.

Sternstein, L. 1976. *Thailand: The Environment of Modernisation*. McGraw-Hill, Sydney.

Vincent, F. 1884. *The Land of the White Elephant: Travels, Adventures, and Discoveries in Burma, Siam, Cambodia, and Cochin-China*. Harper, New York.

Volant des Verquains, J. 1691. *Histoire de la Révolution de Siam arrivée l'an 1688*. J.-C. Malte, Lille.

Wenk, K. 1968. *The Restoration of Thailand Under Rama I, 1782-1809*. University of Arizona Press, Tucson.

Chapter 7

Plus Ça Change . . . Plantations, Tenants, Proletarians or Peasants in Fiji

*R. Gerard Ward**

> . . . the geography of settlement is central, in that villages and cities are rooted in the physical earth but express in their very appearance the spirit of their societies: they link space and society perhaps more intimately than, say, the larger considerations of economic activity on the one hand, the more autonomous activity of thought and art on the other. (Spate 1956: 183.)

> . . . from the air much of the area between Delhi and the Sutlej appears gridded into large squares, with the old irregular boundaries showing up beneath the new. (Spate and Learmonth 1972: 264.)

The rural landscape of an area which has been settled for any length of time is likely to bear the marks of earlier phases of its occupation. Each mode of production, each technology and each society will leave its own distinctive signature, etched in forms appropriate to the particular links it forges between space and society. As subsequent occupiers take over an area they rarely start entirely afresh; elements of the former uses of space are incorporated into the new. One of the

* *R.G. Ward is Professor of Human Geography in the Research School of Pacific Studies, The Australian National University. After taking B.A. and M.A. degrees at The University of New Zealand, he carried out field-work in Fiji in the late 1950s and early 1960s which led to the completion of his doctorate at The University of London. He was Foundation Professor of Geography in The University of Papua New Guinea from 1967 to 1971.*

most persistent settlement elements is the cadastre which, once set out by survey, tends to persist as the basic frame for land occupation even though mode of production or societal form may change. The continuity of the cadastre, particularly in the rectangular grid form, has often been noted (Spate 1956: 178) but mostly in relation to more developed countries. The longevity of Roman land divisions in Westeren Europe (Houston 1953), the continuity of the Jōri system of Japan (Yonekura 1960) and the persistence of rural ownership and field boundaries in the English urban landscape (D. Ward 1962) are notable examples.

Many instances of continuity can also be found in the Pacific islands although there most of the relict landscape features studied so far have been the constructions of the original indigenous inhabitants. Among these are the abandoned taro terraces of many parts of Melanesia, the linear or crescent shaped mounds of old yam gardens in New Caledonia, and the ring-ditch fortifications of Fiji (Parry 1977). Later, decisions about the use and ownership of land had a more significant impact on subsequent rural settlement because they divided customary land (within which traditional systems of occupance have persisted) from government or freehold land (on which non-indigenes have settled and established entirely different occupance patterns). In Fiji this division into different categories of land tenure was laid out in the 1870s and 1880s following the hearings of the Land Claims Commission (Ward 1969) and remains a fundamental element of Fiji's geography.

The landscape of rural settlement in Fiji is thus sharply divided — ethnically, socially, tenurially and economically — between the dispersed homesteads of individual Indian sugar cane farms and the nucleated villages of Fijian rural communities still largely dependent on traditional root crops for subsistence. To some extent this division is also one of physical geography: sugar cane production is now confined to the dry zone of the west. Although commercial sugar farming in Fiji began in the east in the 1870s (Potts 1959) and continued in the Rewa Valley (Fig. 7.1) until the Nausori mill of The Colonial Sugar Refining Company (CSR) was closed in 1959, lower and less economic yields of sugar were obtained than in the drier west.

The present settlement landscape of western Viti Levu and Vanua Levu, with their small cane farms, apparently the reflection of

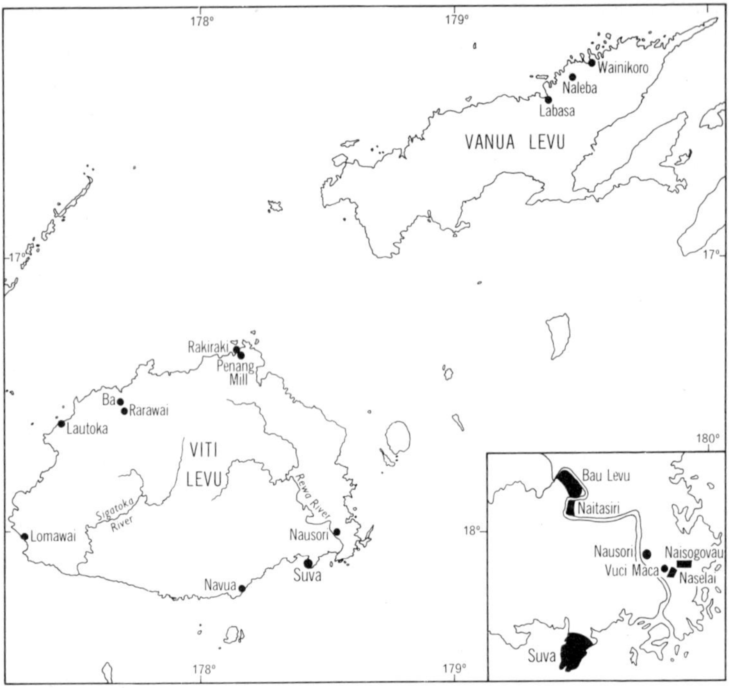

Figure 7.1 Fiji. Places mentioned in Chapter 7.

an independent peasantry, derives in fact directly from an earlier mode of plantation production with indentured workers housed in company 'lines' and working in labour gangs in the large fields of the classical plantation. The lineaments of the old plantations still persist: plantation field boundaries now form tenant farm boundaries, former strip fields appear in aerial photographs as soil marks and old tramway routes are used as footpaths or roads. Similarly, the change from a plantation to a small farm system in the 1910s and 1920s did not reflect as dramatic a change in the mode of production, or in the link between the controllers of society and the land, as might appear at first sight.

After Fiji was ceded to Britain in 1874, the Deed of Cession was interpreted as guaranteeing that land not previously alienated would remain under Fijian control. Claims to prior purchase of land were examined by the Land Claims Commission between 1875 and 1882, and those which were upheld as being based on fair sale, or undisputed occupation, were recognised by the issue of freehold Crown Grants. Although another 104,142 acres (42,146 ha) were alienated in the period 1906-8, the initial 415,296 acres (168,070 ha) of Crown Grants formed the basic area of freehold land on which the sugar plantations were established. Although freehold land (excluding that held by the Crown) covers less than 10 per cent of Fiji it embraces a much higher proportion of the best land, including about 28 per cent of the first class arable land of the three main islands (Ward 1965: 121) and an even higher proportion of the better-drained alluvial flats which were readily accessible from the coast and particularly suitable for sugar cane production.

After the severe rural depression of the early 1870s, brought about primarily by the collapse of the cotton industry, sugar appeared to offer the best prospects for planters on Viti Levu and western Vanua Levu. Many small sugar mills were established (Potts 1959) but most soon closed because of under-capitalisation and labour difficulties. Through the failure or purchase of competitors milling became concentrated in the hands of an Australian-owned enterprise, The Colonial Sugar Refining Company, which opened mills at Nausori (1882), Ba (1896), Labasa (1894) and Lautoka (1903). The Vancouver Fiji Sugar Company operated at Navua until 1922. The Penang Mill at Rakiraki also closed in the same year but was bought by the CSR and reopened in 1927 (Potts 1959).

The primary role of the CSR in Fiji was milling cane and, from the outset, the Company purchased cane from independent planters as well as milling that grown on its own estates. The Company's main aim was to obtain cane as cheaply as possible, and during the 1890s the Company leased many of its estates to European planters, usually former employees, thus divesting itself of much of the direct responsibility for estate labour management.

Land division and layout of estates mirrored the society of the plantations. Control of all operations lay with the planter or the Company's manager. This was exercised through a more privileged group of Indian sirdars who 'were usually persons who looked strong, were senior to others, smart, talkative, capable of organising, authoritative, and could read and write to some extent in English and Hindustani' (Prasad 1974: 17-18). At the bottom of the hierarchy were the indentured workers living in the labour lines, the men receiving one shilling and the women nine pence per day for their task

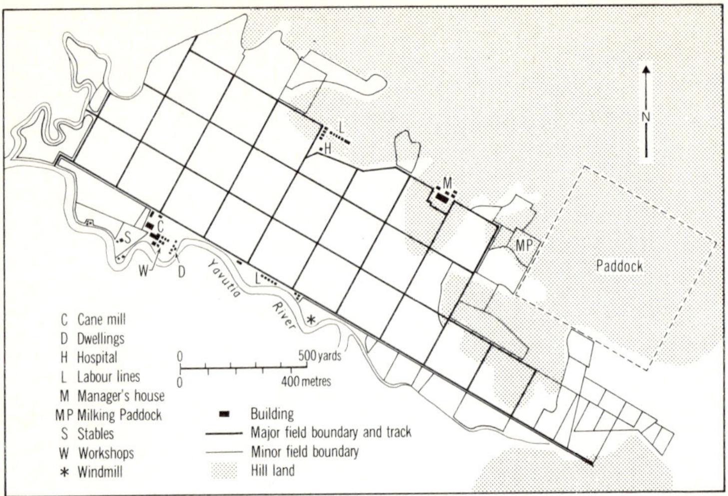

Figure 7.2 Penang Estate, Rakiraki, northeast Viti Levu, about 1910.

of up to nine hours on weekdays and five hours on Saturday (Gillion 1962: 210-11).

Regardless of who owned or managed the larger sugar estates, there was a basic uniformity in their layout and organisation. Cane land was normally assumed to be synonymous with alluvial flats, and before the 1940s very little cane was grown on other than flat land. The flats of each plantation were divided into fields of approximately 40 acres (16 ha), as regularly rectangular as landform allowed (Figs. 7.2 and 7.3). Tracks or tramways separated the fields and provided the means of getting the cane to the mill. Sloping land was used for horse paddocks, and the manager's house — under shade trees — usually occupied a prominent site overlooking the fields. The lines of labourers' quarters were normally placed in some corner unsuitable for cultivation, but not too near the planter's or manager's house. Latrines, hospital, store, stables and tool sheds made up the other principal buildings of the estate. Although the maps on which Figures 7.2 and 7.3 are based were drawn late in the estate period they show clearly the layout of typical estates in the Rakiraki and Labasa areas while Figure 7.6a portrays Naitasiri Estate, on the Rewa, as it was twenty years earlier.

Until the mid-1880s the majority of estate labourers were

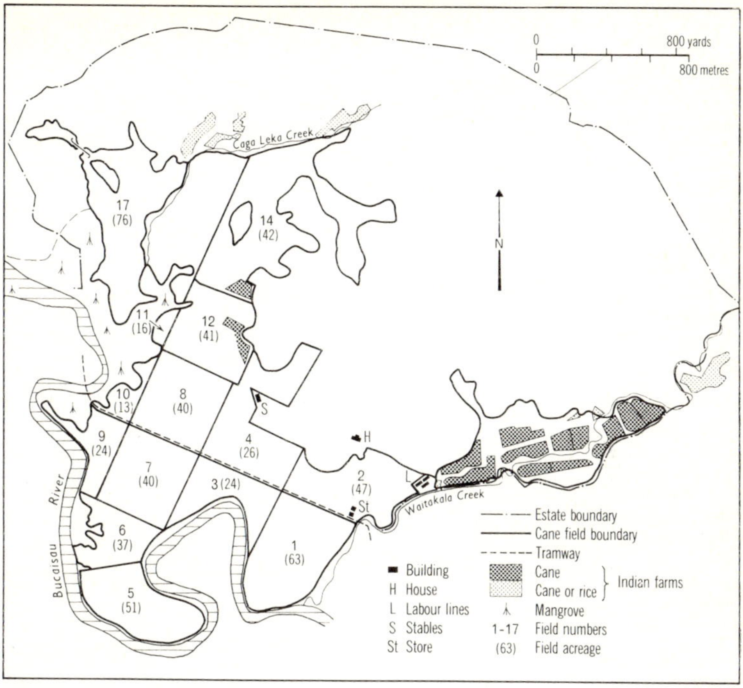

Figure 7.3 Naleba Estate, Labasa, 1913.

recruited from western Melanesia (Ward 1965: 29-31) but thereafter indentured Indian labourers were dominant on the sugar plantations. The labourers worked under indenture which tied them to the estate for five years and then permitted them to seek their own employment for a further five years before being entitled to a return passage to India. Life on the estates was hard, with long hours of work in the fields and poor living conditions in the barracks.

> Each usually had eight rooms on each side, each ten feet by seven feet, or ten feet by twelve feet (after 1908), and assigned to three single men, or to one man, one woman and not more than two children. The rooms had doors but no windows, and for ventilation (but not privacy) the partitions were not carried to the ceiling but were topped with wire netting. There was no floor, except that made by the immigrants themselves out of cow-dung and clay, as in India . . . The lines were crowded, dirty and ugly. (Gillion 1962: 105.)

Work in the fields started at daybreak and continued until the day's tasks had been completed. Labourers worked in men's or women's gangs under the direction of a sirdar and the Company overseer. On independently owned estates the planter was his own

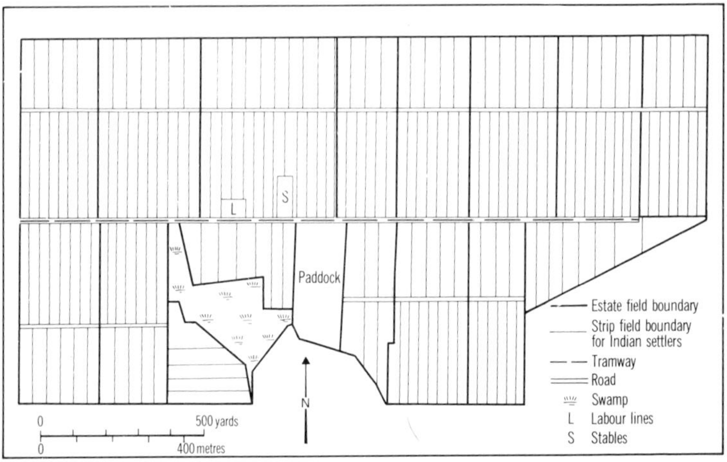

Figure 7.4 Estate fields of Naisogovau, Rewa, subdivided for a 'settlement area' 1918.

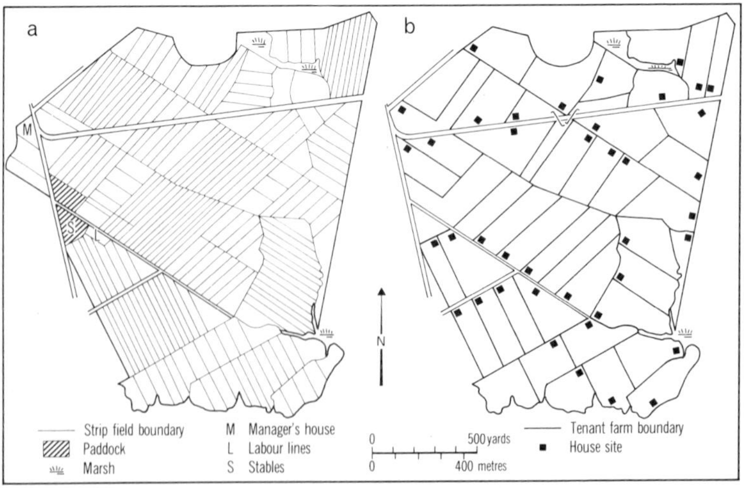

Figure 7.5 Naselai Estate, Rewa, subdivided as a 'settlement area' about 1916 (a), and later as tenant farms about 1929 (b). The latter subdivision is little changed today.

overseer, and one of the advantages to the CSR of leasing land to planters was thought to be that this resulted in closer supervision of labour, higher yields and, for the Company, cheaper cane. Most field work was handwork — planting, weeding and harvesting. The belief that deep ploughing and turning in of trash were necessary to maintain fertility led to ploughing by horse teams or, in some cases, by sets of discs drawn across the field by steam-powered cables. Fertility-maintenance was a prime concern and the cycle of plant and ratoon cane, the ploughing in of cane trash, the growing of green manure crops and the application of fertiliser was rigidly maintained. Field labour requirements on the estates were normally of the order of one man to eight to twelve acres (3 to 5 ha) of cane land so that a labour force of fifty to sixty was not uncommon on larger estates. Once the cane was cut, its transport to the mill by tramway or barge was handled by the Company with a separate labour force. Similarly the mills had their own labour forces. The layout of the fields and of the estates was designed to allow efficiency in control and execution of these operations.

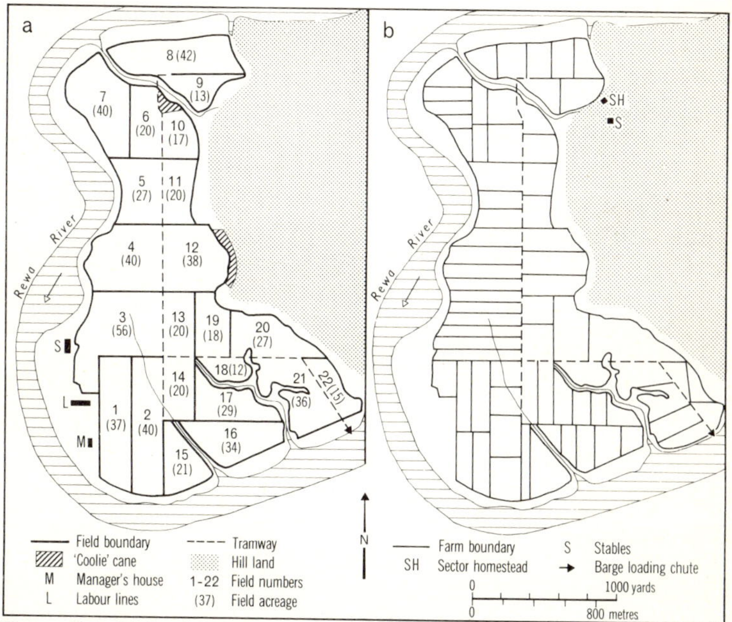

Figure 7.6 Naitasiri Estate, Rewa, 1894 (a), and as subdivided into tenant farms 1929 (b).

When their five years of indenture ended, few labourers were prepared to continue in plantation employment for the second half of their term in Fiji. Planters, anxious to maintain their work-force, saw that an alternative to obtaining further indentured workers (who were not always available) was to employ free labourers as casual workers. Such men might be attracted to a plantation by being given plots of land to cultivate; thus soon after the first indentures ended in 1884 some planters allowed Indians to settle on unused corners of their estates. Figures 7.3 and 7.6a show that on Naleba and Naitasiri estates small areas of cane were cultivated by Indians — the marginal location and small size of these holdings being typical. It was believed that an Indian and his family could cultivate about ten acres (4 ha) without additional labour, but that such a holding would allow the farmer a reasonable livelihood without having to work for casual wages on the estate. Since this would have defeated the planter's goal of establishing a source of casual labour most of the early Indian plots on cane estates were small, rarely exceeding 2 ha. After 1903 the CSR also subleased land outside the cane area to Indians who grew food crops for sale to the Company for the supply of rations to indentured employees (Anderson 1974: 42). Other Indians obtained leases of Crown land in 'Indian Settlements' where holdings, usually of three acres, were used to grow rice and other crops for subsistence and sale to those 'under indenture in some adjacent estate or . . . small town' (Colonial Secretary's Office 1914). For a period between 1910 and 1914 a five acre maximum was imposed on holdings of native land leased by Indians. The reasons for keeping Indian holdings small included a desire to ensure that residents would provide a pool of workers prepared to undertake casual labour for European employers, perhaps at lower rates than would be sought by people dependent solely on wage labour (Anderson 1974: 20-37).

In the early years of the twentieth century opposition to the indenture system increased in India, while better conditions there, and opportunities for employment elsewhere, meant that during the 1900s it was difficult for the Fiji agents to obtain as many recruits as Fiji employers wished. For the CSR this period of labour difficulties coincided with the opening of its large Lautoka mill, the extension of the railway south to the Sigatoka Valley and the opening of new cane land in southwest Viti Levu. At Navua at the turn of the century the

Fiji Sugar Company was purchasing more cane from Indian smallholders than it was growing itself or buying from European planters (Gillion 1962: 100-1).

As early as 1891 the CSR's General Manager, Edward Knox, had advocated 'the settlement of coolies on the plantations' (Archives of Business and Labour [ABL] 1891), and although the idea was not taken up on any scale at that time, the Company did allow some tenants to occupy marginal portions of a number of its estates (Figs. 7.3 and 7.6a). In 1911 the CSR was faced with a possible shortage of 700 to 750 labourers (ABL 1911) and, with the likelihood of a consequent fall in cane supplies from its own or its European tenants' estates, interest in cane grown by Indian tenants increased.

Yields obtained by Indian tenants were usually lower than those achieved on plantations and this was generally ascribed to the difficulty of ensuring deep ploughing because of a lack of horse teams. The Company was also concerned when smallholders grew food crops in place of green manure during the fallow prior to the replanting of new cane, or even grew other crops instead of cane. Lower applications of artificial fertiliser were also thought to be a cause of lower yields and a flaw in smallholder techniques. Thus when the CSR came to experiment more thoroughly with tenant farming the Company sought to devise systems which would overcome these problems.

The first scheme for extensive subdivisions of CSR estates was called the 'settlement area' system. In 1912 it was decided to subdivide land at Vuci Maca, Nausori, for Indian settlement. Single men were to receive a little over three acres and married men more (ABL 1912). The existing division into estate fields was retained but the fields were subdivided into smaller strips. The essential element of the scheme was for each settler to have at least one strip in each of four fields. One field would be in 'plant cane' for harvest, one in 'ratoon cane', one under fallow or green manure and the fourth under newly planted cane. To ensure deep ploughing, CSR company horse teams did the initial ploughing and the settlers were charged a fee for this operation. The remainder of the field work was to be done by the settlers. Figure 7.4 shows Naisogovau Estate as planned for plantation operation and as subdivided under the settlement area scheme. The similarity to a medieval open-field system is striking, and some of the problems of the open-fields soon appeared at Vuci

Maca. Settlers holding adjacent strips did not always maintain similar standards of weeding or field work and on occasion good tenants apparently had to pay for horsework done for bad tenants (ABL 1916a). Some settlers left their holdings when returns proved low: in 1914 the sixty-nine tenants on Vuci Maca grossed only ten shillings per week each (ABL 1915). One of the problems, again reminiscent of the medieval open-fields, was the variability in quality of land, and hence in yields, between strips. Much of Vuci Maca was low and swampy, with difficult soils, and settlers had no additional land for grazing. Holdings were too small and the majority of settlers lived 'a bare existence' (ABL 1916b).

By 1916 the Vuci Maca experiment was recognised as a failure but nevertheless it was proposed to extend the settlement area scheme to Naselai, adjacent to Vuci Maca, with changes to the system of charging for the CSR horsework and more preparation by the Company to bring the land into 'good heart' before settlement (ABL 1916b). The layout of the strips within the old estate fields of Naselai is shown in Figure 7.5a. Naselai, like Vuci Maca and Naisogovau, was a poorly drained area which European planters had been reluctant to take up. The unfavourable soil conditions and the difficulty of maintaining even standards of cultivation across several strips eventually led to much of the field work being done by the settlers working together in gangs. The individual was then 'debited with the wages paid to the members of the gang for work done on his holding' (Dixon 1922b: 4). The individual interest in a holding, and the possibility of higher returns from personal effort were thus lost. Serious discrepancies then arose when end-of-season payments were made to the 'settlers' and not to estate workers who were in effect now doing similar work. Taken together, these disadvantages were such as to lead to the abandonment of the whole system (Dixon 1922b: 5). In mid-1921 it was announced in Lautoka that 'the settlement system is at an end, and all future work on Drasa, Vitogo, Lovu, Lautoka and Lomawai will be on the same wage system as on all other estates' (ABL 1921a). In the case of Naselai, when the tenancies were due to expire at the end of 1922 it was suggested that 'certain marginal fields' be allotted 'on straight-out leases for cane growing' (ABL 1921b). The attempt at an open-field system had failed, yet the continuity of the former field divisions of the estates was usually retained in the subsequent recombination of strips (Fig. 7.5). Even where subsequent field boundaries did not coincide exactly with the edges of strips, aerial photographs (Hunting Aerosurveys 1951) taken thirty years later still reveal some of the old strip boundaries as soil marks in the middle of larger and later fields.

Meanwhile an alternative form of breaking up the estates was tried in all the CSR's mill areas. By 1917 'many estates' which the Company had leased to European planters had 'fallen back' to the CSR. Other leases were about to expire and were expected to be returned leaving the CSR with seventeen estates totalling about 9,000 acres (3,645 ha) for which to find overseers and labour (Dixon 1917: 1). Although it was then still planned to expand the settlement area program, there were disadvantages: one was the difficulty of finding sufficient tenants and another was the need for the Company still to employ a large labour-force for the ploughing and other horsework on the settlements. Dixon's proposal was to subdivide at least one estate in each mill area into farms of about sixty acres (24 ha) to be leased to Indian farmers 'who are known to be men of superior abilities and capable of managing such enterprises' (Dixon 1917: 2). Tenants would be required 'to follow an approved rotation, with ploughing in of trash and green manure etc. in order to maintain the fertility of the land' and to allow this the Company would provide 'full powered teams and the necessary implements . . . the tenant paying for them out of the proceeds of his crops' (Dixon 1917: 3).

The European planter who had held Naitasiri Estate, on the upper Rewa, for some years (Fig. 7.6a) surrendered the lease to the Company in 1917. Later that year a survey was made to subdivide the estate into holdings of sixty to eighty acres (24-32 ha) for the first leases to 'Indian planters' (Potts n.d.: 3). As with the settlement area schemes, the basic integrity of the estate field division was maintained when subdividing for these Indian planters. In most cases a planter's holding could be formed easily from two of the existing estate fields (Fig. 7.6a).

Initially the Company had difficulty finding settlers to occupy these larger holdings, but gradually blocks were taken up in all the mill areas. It was soon found, contrary to expectations, that the Indian planters, like the Europeans farming larger estates, had difficulty finding sufficient labour. An eighty acre holding would require about six to eight workers in about the same ratio to area as on the estates. Furthermore, Dixon noted that 'very close supervision by the Company's Overseers' seemed necessary for successful production, and he concluded that given the difficulties, the area leased to Indian planters in larger farms would not 'represent a large proportion of the total' (Dixon 1923: 4).

The conclusion that the Indian planter scheme would not provide an alternative to cane production on European-managed estates followed closely the abandonment of the settlement area scheme. It also coincided with the period of most serious labour problems faced by the sugar growers and millers. Recruitment of indentured workers from India had been suspended in 1916 and on 1 January 1920 all the remaining indentures were cancelled. Employers were thus faced with the need to restructure their work-forces. In the same month a strike broke out in the Suva-Rewa area and lasted for over a month. In February 1921 a further strike began in northwest Viti Levu, in the core of the cane area, and lasted for six months, seriously disrupting the sugar industry (Gillion 1977: 53ff.). At the time the strikers' demands were viewed by the CSR as 'quite fantastic' (Dixon 1925: 4). The strike finally ended in part at least because Indian cane growers on Company land saw 'that the harvesting of their crops was in danger' (Dixon 1925: 4). A convergence of interest between the Indian growers and the millers was demonstrated.

Recovery from the strike, however, was delayed when world sugar prices fell suddenly late in 1921. The Company announced in February 1922 that the price of cane for 1922 would fall from 20 to 10 shillings per ton and that wages for labourers would be cut from 2s 6d to 1s 6d per day. The estates lost most of their remaining labour-force and many Indian growers did not plant cane for the new season. In April 1923, with some prospect of higher sugar prices under the Imperial Preferential Tariff, the CSR raised wages, offered a bonus on cane prices, and agreed to provide food and other supplies to Indian employees through its stores at prices similar to those existing in the immediate pre-war years (Dixon 1925: 7-9). The result was renewed planting of cane and an increase in the number of Indians seeking work.

Although prospects for the millers now looked better, the CSR was faced with the facts that estates of its own, or of European planters, were scarcely viable, that the settlement area scheme had failed, and that only a small number of Indian planters under the sixty to eighty acre holding scheme would be likely to succeed. Yet the piecemeal leasing of small plots of land marginal to the main plantation fields (Figs. 7.3 and 7.6a) had continued over the years and many of these farmers had been reasonably successful. Other forms of leasing had also been used. In the Sigatoka Valley the Company had subleased 'considerable areas of uncleared land' to Indians for ten years. These 'clearing leases' contained provisions to prevent further subleasing but it was 'found that this clause is evaded in practically every instance, and the greater part, if not the whole of the areas held under these leases were parcelled out amongst sub tenants' (Dixon 1922b: 2). At Navua the Fiji Sugar Company had managed to obtain adequate cane from Indian smallholders from the mid-1890s (Anderson 1974: 49) and from time to time the CSR's European tenants had been given permission to sublease parts of their estates to Indians. The size of these subleases varied considerably, some being four or five acres (ABL 1921c) and others having limits such as 'not exceeding 15 acres each' imposed by the Company (ABL 1920). It appears from the documents available that wherever the leases formed part of existing estates, the estate fields were the basic units for leasing or further subdivision.

By 1918 the CSR's sources of cane for its mills were very varied as can be seen from Table 7.1 which sets out the acreages cultivated in the Lautoka and Rarawai (Ba) mill areas. In the Rewa area, about one-third (40,000 tons) of the cane milled was produced by Indian growers. At Labasa much less cane was grown by Indians. Those on CSR settlement areas provided about 4,000 to 5,000 tons and other Indian growers about 3,000 tons out of the total 80,000 tons milled.

Table 7.1. Areas under cultivation in Lautoka and Rarawai mill areas, Fiji, 1918 (acres)

Sources of cane	Lautoka	Rarawai
CSR's own cultivation	4,450	1,900
CSR's European tenants	5,650	8,300
Other Europeans	4,650*	5,100
Indian tenants on CSR Settlement Areas	2,950	700
Other Indian tenants of CSR	1,450	
Other Indian growers	7,850	*c.* 6,000
Total	27,000	22,000

* Includes 'sundry small growers'.
Source: Dixon 1918.

It seemed that in the Labasa area Indian smallholders preferred to grow crops other than cane although steps were in hand to subdivide 600 acres (240 ha) of the Wainikoro estate into another settlement area scheme and 'to cut up 650 acres at Nagigi into 50/60 acre farms' (Dixon 1918).

By the time the labour crisis was at its peak in 1921 and 1922 the CSR had experience of obtaining cane from Indian growers operating under a wide range of tenurial and farm size conditions. The general failure of the settlement area scheme and the larger Indian planter scheme could be balanced against the experience of those Indian growers who had holdings of about ten acres (4 ha) which they could manage with family labour. Dixon (1923: 1) felt that the Company's experience with such tenants had been fairly satisfactory. It was recognised that smaller holdings were inadequate as a grower with, say, four acres would probably harvest cane from it all for two years but then be faced with having the whole of his holding in fallow or under immature plant cane for the two subsequent years (ABL 1925a). By November 1922 it was certainly clear to the CSR's Fiji Inspector, W. P. Dixon, that

> Our salvation . . . will lie mainly in the small Indian cane-grower . . . there is every reason to believe that we shall be able to settle small farmers on the Estates to whatever extent may be necessary and thus be in a position to concentrate the wage-workers on a much reduced area which we can work intensively. (Dixon 1922a: 1-2.)

The Company acted quickly and began a program of estate subdivision in all its mill areas which was substantially complete by the late 1920s. At first one of the strategies was to subdivide and lease the more isolated estates and the marginal fields of those estates closer to the mills, thus keeping a core of the better cane land of the most accessible estates under Company management. Each estate would be reduced to about 300 acres (120 ha). The subdivision was done in groups of fields as for example when Lord, the manager at Nausori, reported that he was leasing 'only boundary fields at present. Three fields (total 80 acres) abutting on Indian lands at Manoca have now been leased, also one field (20 acres) adjoining Indian at Lakena' (ABL 1924). On Naitasiri Estate, Fields 8 and 9 (Fig. 7.6a) were to be subdivided first into five acre blocks as a special case 'for the benefit of the Bau Levu labour supply across the river' but the remainder of the estate was to be subdivided into ten to twelve acre blocks (ABL 1925b). Mill managers were advised that 'areas leased should be as much as a man can properly handle', the earlier policy of endeavouring 'to create a labour supply for the estate by giving Indians less area than would occupy their full time' having now 'served its purpose'. Leases should now be 'from 8 to 10 acres according to the class of land' (ABL 1927). In planning future leasing managers were advised that blocks should 'be arranged to the best advantage considering all factors such as boundaries, drainage, harvesting, house sites, etc.', and 'that each farm should be in one block rather than that the farmers have interest in two or three fields' (ABL 1925b).

Progress in subdivision and leasing was rapid. In 1925 some 10 per cent of the cane area was cultivated by Indian and Fijian tenants of the CSR and 52 per cent by the Company itself. By 1930 the equivalent figures were 36 and 22 per cent, and in 1933 the tenants were cultivating 43,077 acres (17,433 ha) or 51 per cent of the area under cultivation, and the Company only 7,450 acres (3,015 ha) or 9 per cent (Shephard 1945: 38). Thereafter the area in estates fell still further and in effect the only land cultivated by the Company was maintained primarily for experimental purposes.

The new landscape of small tenant farms and individual homesteads appeared to represent the visual and spatial expression of the transformation from use of a proletarian wage labour force to peasant production. But just as the new farm boundaries tended to represent subdivisions of the estate fields, thus maintaining continuity in land division, so the conditions of the tenancies maintained much of the Company's control over the tenants and the farming operations. The Company's primary concern was still the maintenance of supplies of cane for its mills, and its managers still believed that Indian farmers, without supervision, would not be able to maintain soil fertility and yields. Not only would this jeopardise the volume of cane supplies but it might also lead to demands for higher cane prices. In 1921 Dixon had noted apropos of this risk that 'we have it in our power to make such conditions as we think fit for the fallowing, conservation of organic matter in and manuring of the soil, depth of ploughing etc.' (Dixon 1921: 3). The Company's goals were achieved through the exercise of this power by imposing conditions of tenancy which prohibited subdivision and subleasing, laid down prescribed rotations of cane and green manure crops, prohibited the growing of rice on caneland during the fallow period and required the tenant to take direction in harvesting from the Company's field officers. The area supplying each mill was subdivided into 'sectors', some of which coincided with former estates or groups of estates. The sector field officer maintained checks on drainage conditions, organised work groups (for which the tenants were debited), set the harvesting schedule and ensured that the farmers were organised into harvesting gangs for group work. The Company also organised the supply of fertiliser to the tenants and the tenancy agreement required the grower to apply the fertiliser as directed. The cane purchase agreements required the farmer to plant by a certain date and specified the area of cane which would be purchased in any one year. The Company retained rights-of-way over a farm for removal of cane from neighbouring areas and the grower was prohibited from giving any lien on his crops to other than the Company. The various agreements were developed in the 1920s and 1930s, and those which were in use in the early 1940s clearly demonstrate the tight control and supervision which the Company was able to maintain (Shephard 1945: 51-8). There is no doubt that this level of supervision was a key element in the maintenance of productivity on the tenant farms.

The present landscape of Fiji's sugar zone is sometimes assumed to be a new creation following the labour problems of the early 1920s. Closer examination shows that the break with the past was not as sharp as outward appearances might suggest. It is clear that the farm settlement pattern was a direct derivative of the earlier estate pattern. Similarly the new tenants were still not so much independent peasant farmers as semi-proletarian farm workers whose productive activities remained under close control of the sugar milling company from which they held their tenancies.

Acknowledgments

Much of the information and particularly the maps on which this essay is based were collected in Fiji in the early 1960s, while working under a grant provided by the Colonial Social Science Research Council, from the various offices of The Colonial Sugar Refining Company. I am particularly indebted to Mr J. C. Potts, then the Company's General Manager for Fiji. Through his kind offices and with the help of the mill managers, I was able to see and copy the original maps on which the line drawings are based. These maps — of which there is no catalogue — were then kept at the following mills: Figure 7.2 at Penang, Rakiraki; Figure 7.3 at Labasa; and Figures 7.4, 7.5 and 7.6 at Nausori. More recently I have had access to the CSR records now held in the Archives of Business and Labour at The Australian National University, Canberra. I am grateful to the CSR for permission to use the Company records, and to Maureen Purtell of the ABL for her help. Nancy Clark, Department of Human Geography, assisted in my use of archival sources and in preparing the paper. Many years ago, Associate Professor J. S. Whitelaw, then a doctoral student at The Australian National University working in Suva, checked a number of factual matters on my behalf. Associate Professor A. G. Anderson also assisted by making material available to me. I would like to express my gratitude to all these people and organisations.

References

Anderson, A. G. 1974. *Indo-Fijian Smallfarming*. Auckland University Press/Oxford University Press, Auckland.

Archives of Business and Labour (ABL) 1891. 142/2610: 92-3. 23/12/1891, Rarawai Mill Out.

Archives of Business and Labour 1911. 142/3129, 21/3/1911, Nausori from H/O Private.

Archives of Business and Labour 1912. 142/2267, 6/8/1912, Nausori Mill In.

Archives of Business and Labour 1915. 142/2269, 2/3/1915, Nausori Mill In.

Archives of Business and Labour 1916a. 142/2270, 14/3/1916, Nausori Mill In.

Archives of Business and Labour 1916b. 142/1061, n.d./3/1916, Private Out.

Archives of Business and Labour 1920. 142/3394, 30/4/1920, Fiji, Rarawai, European Tenants.

Archives of Business and Labour 1921a. 142/2923, 4/7/1921, Fiji Inspector (W. P. Dixon) In.

Archives of Business and Labour 1921b. 142/2273, 19/9/1921, Nausori Mill In.

Archives of Business and Labour 1921c. 142/3397, 20/12/1921, Lautoka, European Tenants.

Archives of Business and Labour 1924. 142/3139, 11/4/1924, Nausori to H/O Private.

Archives of Business and Labour 1925a. 142/3158, 17/6/1925, Fiji Inspectors' Notes for Managers.

Archives of Business and Labour 1925b. 142/3158, 11/11/25, Fiji Inspectors' Notes for Managers.

Archives of Business and Labour 1927. 142/3158, 17/5/27, Fiji Inspectors' Notes for Managers.

Colonial Secretary's Office 1914. Commissioner of Lands to Colonial Secretary, 18 May 1914, CSO Minute Paper 3680/4, Fiji Archives.

Dixon, W. P. 1917. Leasing to Indian tenant farmers. TS. CSR Archives, Sydney 10/10/17.

Dixon, W. P. 1918. Notes on position in Fiji. TS. CSR Archives, Sydney.

Dixon, W. P. 1921. Our Fiji business. TS. CSR Archives, Sydney 2/9/21.

Dixon, W. P. 1922a. Future policy in Fiji. MS. ABL 142/2923, 9/11/22 Fiji Inspector (W. P. Dixon) In.

Dixon, W. P. 1922b. Various types of leases to Indians. TS. ABL 142/2923, 22/2/22 Fiji Inspector (W. P. Dixon) In.

Dixon, W. P. 1923. Indian cane farmers. TS. ABL 142/2923, 17/1/23 Fiji Inspector (W. P. Dixon) In.

Dixon, W. P. 1925. Crisis in the Fiji sugar industry 1920/1925. TS. CSR Archives, Sydney 6/7/25.

Gillion, K. L. 1962. *Fiji's Indian Migrants: A History to the End of Indenture in 1920.* Oxford University Press, Melbourne.

Gillion, K. L. 1977. *The Fiji Indians: Challenge to European Dominance 1920-1946.* Australian National University Press, Canberra.

Houston, J. M. 1953. *A Social Geography of Europe.* Duckworth, London.

Hunting Aerosurveys 1951. Aerial photographs of Fiji Run 20/I.

Parry, J. T. 1977. *Ring-ditch Fortifications.* Bulletin of the Fiji Museum No. 3 [Suva].

Potts, J. C. 1959. The sugar industry in Fiji, its beginnings and development. *Transactions and Proceedings of the Fiji Society* 7: 104-30.

Potts, J. C. n.d. [1961?]. Sugar cane farming in Fiji, the six phases of development. Mimeo, 4 pp.

Prasad, S. 1974. *Indian Indentured Workers in Fiji.* South Pacific Social Sciences Association, Suva.

Shephard, C. Y. 1945. *The Sugar Industry of Fiji.* His Majesty's Stationery Office, London.

Spate, O. H. K. 1956. Bush and city: some reflections on the Australian cultural landscape. *Australian Journal of Science* 18: 177-84.

Spate, O. H. K. and Learmonth, A. T. A. 1972. *India and Pakistan: Land, People and Economy.* Methuen, London.

Ward, D. 1962. The pre-urban cadastre and the urban pattern of Leeds. *Annals Association of American Geography* 52: 150-66.

Ward, R. G. 1965. *Land Use and Population in Fiji.* Her Majesty's Stationery Office, London.

Ward, R. G. 1969. Land use and land alienation in Fiji to 1885. *Journal of Pacific History* 4: 3-25.

Yonekura, J. 1960. *Comparative Studies of Rural Villages between Ancient Japan and China.* Kokon Shoin, Tokyo. Quoted pp. 146-50 in K. Fuyioka, Development of historical geography in Japan. *Japanese Geography 1966,* Association of Japanese Geographers Special Publication No. 1, 1966.

Chapter 8

Tropical Mountain Forest in Retrospect and Prospect

*Geoffrey Hope**

Eighteen hundred years ago one of the greatest of the armchair geographers, Claudius Ptolemy, wrote of the 'Mountain of the Moon, whose snows feed the lakes, sources of the Nile' (Schlichter 1891). Although Ptolemy's view of the nature of the lunar mountains probably differed from the bleak and lifeless crags revealed on television, in one sense the name was prophetic; many areas on the equatorial mountains of East Africa have lost a forest cover and are taking on a more desolate appearance as soil is lost and the catchments are degraded. This is also true in many other tropical areas.

Extensive areas of grassland or open woodland occur on many of the African, American and Asian tropical mountains. Some of these areas seem to be 'natural'; they have a rich, well adapted, herbaceous flora and no evidence that they have replaced mountain forests. Others show clear signs of having been recently produced from forest by felling or burning, while a third group cannot clearly be allocated to 'natural' or 'anthropogenic'. The causes and controls of these openings in the forest have created considerable investigation and

* *G.S. Hope is a Lecturer in the Department of Geography, School of General Studies, The Australian National University. Living next door in 1953, he was commonly underfoot in the Spate household. He joined the Department of Biogeography and Geomorphology, Research School of Pacific Studies, in 1969 and carried out research into the vegetation history of the high New Guinea mountains while holding doctoral and post-doctoral awards.*

debate; however, while this proceeds, the openings are remorselessly being widened to such an extent that the forest is being reduced to islands in a sea of grass. To appreciate the scale and effects of changes in progress, or planned, for the productive zone between 1,000 and 3,000 m in the tropics, some understanding is needed of the changes that forests in this zone have experienced in an undisturbed state and under various regimes of pre-industrial utilisation.

Vegetation histories (based on fossil plant material, carbonised particle frequencies, radiocarbon dating and archaeological evidence) can help to unravel the status of the existing vegetation and can provide estimates of the rate of clearance of, or invasion by, forest in the past. The interpretation of vegetation histories can leave considerable doubt about the cause of vegetation change, but the fact of the change, and the response to it by the component communities can tell us much about the properties of the communities and the ability of individual species to resist or exploit change. These data often cannot be gathered from short-term observations of modern situations, yet they can be of great assistance in formulating management decisions.

Tropical Mountain Forests and Grasslands

Vegetation historians have tended to favour a climatic explanation for the existence of grassland in the montane tropics. In tropical South America, Van de Hammen (1974) reviewed evidence for periods when grassland replaced montane or lowland forest, and demonstrated that this process has been repeated several times during the last million years. The times of forest cover are held to reflect warm, relatively dry, conditions and those of grassland steppe indicate cool, relatively more humid periods. This may be straightforward for the large swings in climate to and from glaciation, but climate should not be regarded as the absolute control of all changes over the last 12,000 years when most areas became increasingly forested, while a few remained unchanged or lost existing woodland. The tropical Andes in many areas today are completely stripped of timber and are used for grazing or potato growing. If a climatic explanation is rejected, clearance may have started in some areas about 10,000 years ago; perhaps in some cases the forest was never cleared but simply prevented from returning despite a newly favourable climate at the end of the Ice Age.

In East Africa there are several vegetation histories from mid-montane altitudes, many of which demonstrate unequivocal clearance followed by a failure of the forest to reinvade. The Virunga volcanoes of the Congo have been extensively stripped of forest, and parts are now almost useless for grazing because of erosion brought about by goats. The controls on the savannah and lower montane forests are still a subject for argument between proponents of climate, grazing (by elephant or giraffe) or fire as the principal agent. Because of the very long history of man and fire in East Africa, vegetation histories covering only the last 50,000 years do not help us to choose the dominant influence: probably nowhere has absolutely undisturbed vegetation developed during the persistence of the present climatic regime of the last 10,000 years. The vegetation histories available from the mountains (reviewed by Livingstone 1975) show a general shift from grassland to forest around 9,000 B.P., presumably due to warmer climates. Subsequent changes may have had climatic or anthropogenic causes but clearance for agriculture is apparent about 1,000 years ago.

In New Guinea, the central mountains were at first considered uninhabited: the densely populated grassland and garden areas of the intermontane basins came as a surprise to the first European explorers in the 1930s when they had penetrated the thinly populated tropical and montane forests of the outer flanks of the mountains. The discovery was repeated at many locations along the central ranges: wherever soils are of reasonable fertility and valleys are broad enough to generate a local climate with sunny afternoons, the forest is absent and grassland or vegetable gardens occupy slopes and swampy valley floors. Forest is restricted to isolated patches in the settled areas, although it is more widespread on high surrounding ridges. It is apparent that the landscape is favourable for tree growth, for everywhere are planted groves of *Casuarina oligodon.* The presumption that much of the area was formerly occupied by forest is demonstrated by vegetation histories (e.g. Powell 1970; Powell *et al.* 1975; Walker 1970; Walker and Flenley 1979) which show that forest cover had once been complete and that clearance started at different times after 10,000 B.P. These highland non-forest areas total only 12,000 km^2 compared with over 180,000 km^2 of montane forest still extant in New Guinea. Forest cover is almost complete on the outer flanks of the mountains where settlement is scattered or absent.

By contrast, the hill forests of Java, Sumatra, Borneo and mainland Southeast Asia are now extensively cleared. As in New Guinea, forest is recognised to have formerly covered all the montane area, but it is not known when clearance began except in central Sumatra, where Flenley *et al.* (1976) have suggested that it had already started 8,000 years ago.

Australia provides a complete contrast to the Asian and New Guinean picture. Tropical montane forest is only found on the wetter mountains of northeast Queensland. Although fire is acknowledged to be a major influence in the drier *Eucalyptus* forests and woodlands, it seldom penetrates undisturbed rainforest, although it may determine the exact location of the boundary between the communities (Singh *et al.* in press). Although the eucalypt savannah/rainforest boundary is convoluted and rather diffuse, few 'natural' or man-made grassland areas occurred historically within the jungle area. Vegetation histories (e.g. Kershaw 1976) show that the tropical forest of northeast Queensland has not changed radically over the last 7,000 years but that the influence of fire was of considerable importance during drier conditions prior to that time. The absence of distinct montane grassland, despite the rather marginal situation of the montane forest in tropical Australia, is interesting and it throws some doubt on the 'natural' status of grasslands or thorn-savannah in other tropical montane areas where people have taken a more active role in forest clearance for agriculture or grazing. The Queensland montane forest has, of course, been extensively felled over the last 100 years.

Thus many tropical areas have experienced shifts from forest to non-forest vegetation, sometimes repeatedly during the Pleistocene and Holocene. Clearly, many of the regional results of such a shift will depend on the extent, rate and causes of the removal. The contribution of historical geography is to help us understand the long-term results of a variety of forest disturbance events associated, for example, with pre-agricultural and 'traditional-agricultural' causes. In almost all tropical areas in the world it is impossible, in a strict sense, to investigate the effects of 'natural' clearance on present day montane forests because the imposition of Holocene climates and consequent development of modern forest communities post-dates man's arrival. Places where people came very late, such as Hawaii, also have exceptional forests as a result of the same isolation that excluded man for so long. However, it is possible to consider Pleistocene forest responses in some areas (South America, Melanesia) where there is a relatively short history of human contact. Examples of apparently little-affected forests can also be sought in Africa and Southeast Asia. New Guinea provides examples of Pleistocene pre-human forests and modern 'natural stands' and also displays a wide range of agricultural strategies; thus specific cases will be drawn from there.

The Forest before Man

The montane forest of New Guinea represents a surviving arm of the austral cool temperate rainforest, which was once one of the great forest biomes. Other remnants of this biome occur in New Zealand, New Caledonia and Tasmania and a few taxa still occur in mainland Australia. The floristic relationships between these regions and also with South America, together with late Cretaceous-early Tertiary records of fossil pollen and plants, show that the origins of cool temperate rainforest extend back to the late stage of Gondwana break up, with the sundering of Australia and Antarctica fifty million years ago. The subsequent slow drift northward of the Australian plate provided climates that at first favoured the closed forest which remained widespread in Australia until the middle Miocene. But periods of drier climates since then have restricted the forest more and more on the mainland, and the sclerophyll, fire-resistant Australian vegetation has become important. Even so, the 4.3 million year old Pliocene flora of Grange Burn, in western Victoria, includes the conifers, *Dacrycarpus, Dacrydium* and *Araucaria,* which are important in modern New Guinea montane communities, and phalangers, ringtail possums and bandicoots related to modern New Guinea species emphasise the similarities (Gill 1957). Although the continued northward drift of the continent through the Tertiary and finally the cooler, drier climates of the Pleistocene doomed the forest to virtual extinction in Australia, the drift also provided a refuge when the collision of the Asiatic and Australian plates threw up the extensive mountain cordillera of New Guinea. Of course the forest communities present today are changed from those of the Tertiary forest through adaptation and immigration of new taxa, particularly from Southeast Asia. To the fauna has been added a bewildering array of rodents. Overall, however, it can be concluded that although montane forest is intolerant of drought, even only seasonally applied, it has been selected for elements with sufficient mobility and plasticity to accept slow but massive change.

On a decreasing time-scale, natural causes of forest clearance in New Guinea include climatic change, tectonic-volcanic instability, disease, fire and storm. The process of natural forest change due to climatic stress is not well understood, partly because even the mechanisms by which forest limits are maintained are not agreed upon. Occasional episodes of severe frost down to lower altitudes than usual (as in 1972 in the Papua New Guinea highlands) provide an indication of the processes involved: selected species suffer foliage loss; adaptable species then reshoot and recover, while other species are permanently lost. Climatic deterioration presumably takes the form of an increasing frequency of extreme climatic events, which remove some species immediately and retard others. Eventually only the resistant species are left, and these will usually be those that occur at the upper limits of stable vegetation communities. If the process is slow, the resistant species will have time to spread, and the canopy will remain closed. If change is more rapid, the canopy is opened up and frost effects are then felt within the forest, killing saplings, shrubs and seedlings. The younger stages of even frost-tolerant tree species will be sensitive to the still, cold air in the gaps and the forest will fail to replace itself, leaving isolated mature trees which are liable to storm damage and which eventually join their comrades, fall and rot. Where cold conditions are responsible for forest retreat, the rate of organic breakdown is slowed and results in very slow release of nutrients to the grassland or shrubland that replaces the forest. Examples of this process in the past are known from Sirunki (Fig. 8.1), where between 28,000 and 20,000 B.P. subalpine forest (which today forms a zone above montane forest) was several times replaced by subalpine grasslands, with no apparent change in sediment load to the catchment (Walker and Flenley 1979).

The reverse process, that of reforestation after warmer conditions were re-established, has been described at several sites in New Guinea. At Sirunki, and indeed at other sites below the present upper limit of montane forest, it was the montane forest itself, rather than subalpine forest, which replaced grasslands around

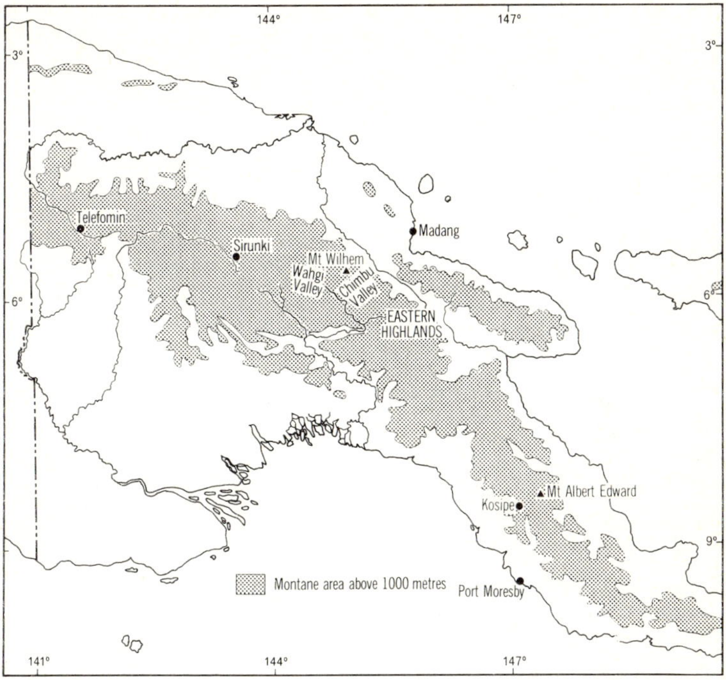

Figure 8.1 The montane area of New Guinea and places mentioned in Chapter 8.

12,000-10,000 years ago. Walker and Flenley estimate that the forest established itself across fifty kilometres in 1,500 years. This rate suggests a very aggressive colonisation by forest that included taxa such as *Nothofagus* which is regarded as very slow to spread at present. The fact that a zone of subalpine forest does not precede the advancing forest shows that reforestation is not the simple reverse of clearance; the arrival of non-limiting climates permits rapid growth even at the forest limits and a consequent selection of fast-growing species as community dominants. Colonisation is assisted by the occurrence of mature soils in the new area and by the absence of disturbance by man. Because of the frost-sensitive nature of seedlings at the forest edge it should be remembered that a given frequency of severe frost at a site (for example once every thirty years) may be insufficient to remove established forest, but enough to prevent recolonisation once the forest has gone. It may thus be misleading to ascribe forest changes directly to changes in the average climatic parameters.

Tectonic and volcanic disturbances have been, and are still, important causes of clearance in New Guinea. Near Madang, Pain and Bowler (1973) conclude that almost all the forest on moderate to steep slopes is affected by earthquake induced landslips. The Madang earthquake of 1970 caused slips over 25 per cent of the area strongly affected. Landslides are so ubiquitous that it is safe to say that no area of montane forest in New Guinea has a completely unbroken canopy. Volcanic eruptions also cause forest loss, largely through burial by airfall ash. The effect of these types of disturbance is continually to provide a reversal of natural succession and consequently maintain a pool of taxa adapted to succession (White 1975). It also breaks up the forest into a mosaic of stands of different maturity and helps to retain diversity.

Disease and insects play a largely unknown role in the montane forest. As particular species are singled out, the usual effect is to change the forest composition rather than cause clearance. By creating openings in the canopy and dry fuel, diseased forest becomes more susceptible to burning. Natural fires are rather unlikely in the montane forest unless the canopy is opened and the humidity lowered. Fires thus tend to occur in forest that is under stress for some reason, and they were particularly common in montane forest during the 1972 drought. Trees damaged by lightning strikes are fairly common, but examples of this leading to forest fire are very rare in New Guinea.

In summary, the last million years in New Guinea have imposed relatively gradual but large and widespread climatic changes and numerous abrupt but restricted disturbances of the montane forest. The result is diverse forest communities with numbers of colonial taxa as well as primary taxa. The distinctive feature of natural clearance is the preservation of mature soils and seed sources at, or near, the clearance site and a low incidence of repeated disturbance at any one site.

The Coming of Man

I once travelled from Tifalmin to Busilmin, in the Eastern Star Mountains of Papua New Guinea. After two hours' walk through grassland the track entered the forest and I did not emerge (except for two river crossings) for four days. This relatively short time was sufficient to affect my attitude to forest. Despite their beauty, the apparently endless trunks and dense gloomy canopy made it a physical relief finally to emerge into open space. I suspect that even pre-agricultural societies widened natural clearings with fire and perhaps by felling without a clear economic incentive. Such clearances would resemble the effects of landslides and would be reversed in due course, except where the forest was at a limit to growth, in a 'climatic tension zone', as at an altitudinal or soil boundary. There the minor clearances could cause a permanent adjustment to contemporary conditions. Minor changes to the forest are not likely to be recorded in vegetation histories, but changes in the rate of burning can be rather sensitively detected by counting fine carbonised particles preserved in stratified sediments.

Clearance for agriculture involves the removal of trees, disturbance of the soil and often a continued effort to restrain the forest from reoccupying the site. In the simplest case, only one crop is retrieved and the forest is allowed or encouraged to re-establish itself, the process differing from landslide damage in that bare ground is exposed longer and that positive steps are often taken to ensure good drainage, with consequent leaching (Clarke 1976). Also all trees are usually not removed, very large, very hard or economically valuable trees (for example *Castanopsis* and *Lithocarpus,* acorn producing taxa) being left behind. Some crops such as taro (*Colocasia esculenta*) can be grown beneath a tree canopy after relatively light clearing.

From this type of minimal disturbance there is a continuum to intensive sedentary agriculture with sophisticated methods of replenishing soil fertility and maintaining appropriate soil textures. Vegetation histories and archaeological research provide case studies of the effects on mountain forest of four agricultural regimes.

Kosipe

Kosipe Mission, at 1,950 m altitude in the Owen Stanley Range, is the site of one of the oldest records of man in New Guinea (White *et al.* 1970). Evidence for occupation comes from stone tools and hearths which date from about 26,000 B.P. to perhaps 8,000 B.P., but at European contact the area was uninhabited and supported closed forest. Preliminary carbonised particle analyses (Hope, unpublished data) from an adjacent swamp show the appearance of significant burning about 30,000 years ago which later rose to high levels and then returned to very low intensities at some time after 9,000 B.P. The vegetation history indicates little discernible change in forest composition from 35,000 to 10,000 B.P. (although local changes on the swamp indicate the onset of colder conditions around 20,000 B.P.) and the sediments indicate no substantial loss of soil from the surrounding slopes. There was a change at about 10,000 B.P. from beech *(Nothofagus)* forest to mixed oak forest but this most probably reflects a climatic warming. A decline in oak and concomitant increase in other montane trees took place at the same time as the cessation of burning and the possibility exists that this was a result of abandonment of partly cleared areas. For the Pleistocene occupation period at least, pre-agricultural man seems to have had little effect on the forest, despite the use of fire.

Wahgi Valley — Kuk

The Kuk area, at 1,600 m near Mount Hagen in the central highlands, is currently accepted as having experienced a 9,000 year history of continuous dry land agriculture, with several phases of drainage and utilisation of swampy plains (Golson and Hughes 1978). Pollen analyses and dating of many sections in the region (Powell *et al.* 1975) have established a general picture of vegetational change. As at Kosipe, a beech forest occupied some of the area from before 25,000 B.P. until about 12,000 B.P., and no anthropogenic effects can be discerned, although traces of early man are known. Many sediment sections are truncated or disturbed after 12,000 B.P., and low-lying areas preserve a thick clay layer, first deposited about 9,000 years ago. Where pollen records exist, they indicate that the forest had been effectively cleared by 5,000 B.P. and probably earlier. Remnant tree taxa indicate a shift to species associated with *Castanopsis*-mixed forest, as at Kosipe. Golson and Hughes (1979) interpret the widespread clay layer as resulting from at least a threefold increase in the rate of erosion of the slopes between 9,000 and 6,000 B.P., consequent on forest clearance. Agricultural phases are indicated during this time and subsequently by several sets of drainage ditches, some of which became infilled and imply temporary abandonment of the swamp floor. The pollen results only cover the last 5,000 years, but show that montane forest has been unable to re-enter the catchment, and has generally been as restricted as at present. The Kuk basin floor is not used for traditional agriculture today, but a large population farms the nearby slopes. It is notable that many pioneer tree species have been present throughout the history of the site, expanding at times of disuse. Montane forest has thus maintained a foothold in the region through a very long period of disturbance.

Telefomin

In an interesting contrast the impact of man on forest areas at Telefomin (1,200 m), although probably of shorter duration and lower intensity, seems to have resulted in a more permanent opening up of the forest. A vegetation history for the last 20,000 years has been obtained from a cleared area close to a permanent settlement (Hope, unpublished). The record indicates that a closed swamp forest occupied the site until after 8,000 B.P. Unlike vegetation histories from higher altitudes in New Guinea there were no pronounced changes in forest composition at the time of climatic warming about 12,000 to 10,000 years ago. Sedimentation was interrupted between 8,000 and about 4,000 years ago, by which time partial clearance had taken place. This break in the core appears to be analogous to those noted in many of the Wahgi Valley cores, when erosion occurs during clearance, removing some of the pre-clearance record. Increasingly open conditions are indicated up to the present day, even though the site has never been cultivated in living memory.

About 25 km^2 of steep hillslopes and boggy gentle fan surfaces are almost completely devoid of trees, even though they are now never gardened. Most agriculture undertaken by the relatively sparse population involves opening up new plots in montane forest, often necessitating travelling long distances from their houses. Similar grasslands occur around the Strickland River gorge and in the Asaro Valley but their age is unknown. The completeness of the deforestation in the eastern parts of the central highlands led Robbins (1960) to speculate that settlement must have occurred much earlier there than in the western parts, in places such as Kuk. Since the fossil evidence contradicts this correlation of degree of clearance to length of settlement, other explanations must be sought. As the Asaro region has relatively low rainfall (1,400 mm) and a dry season, it has been suggested that the easily fired grasslands are so often burnt that forest is excluded. The Strickland and, particularly, Telefomin grassland areas have very high rainfall (>4,500 mm) and are less readily burnt. There it is loss in soil fertility after clearance which seems both to prevent continued gardening and to hinder re-invasion by forest. It is possible that changed soil texture and fertility have a role in the Asaro area also.

Upper Chimbu Valley

The Chimbu Valley, on the southern side of the Bismarck Range, is today one of the most heavily populated areas in New Guinea, with densities of up to 75 persons/km^2 compared with Asaro Valley populations of 15-25/km^2 and 5-10/km^2 at Telefomin. As in the Wahgi-Kuk case, the settled areas support much tree growth, because a tree fallow including several species from the montane forest, but principally the nitrogen-fixing tree, *Casuarina oligodon,* is used to restore garden fertility. A few large trees of *Araucaria huntsteinii* and many stands of other trees are preserved as having ritual significance. Erosion of soil from gardened slopes is recognised as a problem, and efforts are made to control slopewash with barriers. Increasing population has probably shortened fallow times, led to migration onto the northern slopes of the Bismarck Range, and encouraged clearance of new plots in primary forest at ever higher altitudes, currently up to 2,700 m. Records kept by the missionaries who entered the valley in 1933 indicate that in one area the forest edge then lay about two to five kilometres downslope from its position in 1973. Vegetation histories upvalley from the garden areas (Hope 1976) first show evidence for clearance of montane forest at about 1,200 B.P., and the introduction of *Casuarina* at 1,000-800 B.P. The rate of clearance, and of *Casuarina* planting, seems to have accelerated after about 300 B.P. This presumably reflected the approach of settlement closer to the pollen analysis sites and was probably the result of an expansion of population. This may have been due to the introduction of sweet potato to the highlands, which would have allowed cultivation at higher altitudes. Since the Chimbu pollen analysis sites do not register the early intensive agriculture taking place in the Wahgi Valley, these results probably reflect changes within twenty kilometres of the site. Clearance rates in the last 1,000 years were thus probably in the order of 1 km/century, rising to 3-4 km/century in the last 300 years. The historical rate of 5 km/century probably exceeds any pre-contact rates due to pacification and the widespread introduction of the steel axe.

The Upper Chimbu situation then, although appearing to be as well adjusted as the Wahgi Valley floor to settlement, represents relatively few generations of cropping and increasingly intensive land use. The nutrient and soil 'capital' built up by the previous forest may still be being spent, although the importation of techniques developed to maintain fertility in longer-settled areas may be able to delay or prevent the onset of crop failures. The silviculture of *Casuarina* and other trees is the most important technique, and *Casuarina* provides substitutes for many forest products no longer accessible as the montane forest boundary retreats.

Current Prospect for Montane Forest

Increase in population, the introduction of cash crops such as coffee and the need for grazing for beef production have greatly increased the demand for cleared land in the montane zone. At the same time local demand for forest products, principally firewood, housing materials and fencing pickets, but including bridgework timber, has increased. Additionally, a new demand for sawn timber has been imposed by local and national needs, and in a few areas the export log, timber and pulp trade has an interest in montane areas. (Most export logging in New Guinea is currently restricted to lowland forest close to port facilities.) White (1976, 1977) has defined forest resource management guidelines:

(a) timber or other forest production management should aim to maintain the potential for maximum sustained yields;
(b) areas too steep or unstable for timber extraction should remain as protection forest and be managed for soil protection, water regulation and other values;
(c) forests should be managed so that composition and structure are altered no further than is necessary from those of natural forests;
(d) operations should cover relatively small annual coupes with due care taken to limiting change in any one ecosystem in a working plan over a time period;
(e) operations should provide a mosaic of coupes of varying stage (structure) and ages in each ecosystem;
(f) adequate areas of native vegetation should be retained, representing the range of environmental conditions present in the block;

(g) native vegetation should be kept in strips along perennial watercourses and risks of stream pollution and erosion minimised;
(h) fire should be excluded from the forest;
(i) government and community backing are needed by forest managers to protect the forest from external disturbance, e.g. fires and illegal felling.

White (1977) concludes that New Guinea forests are not as commercially attractive as other tropical areas because of variable forest conditions and the logistics of extraction, but he believes that world timber shortages will eventually force their exploitation.

Routley and Routley (1977) are doubtful whether any responsible management guidelines have been followed in lowland forestry, where such management practices might in any way reduce the profitability of the initial harvest of existing forest. Their economic analysis shows that the costs of small patch cutting and selective logging, as recommended in the guidelines, are considered too high by private companies and that clear felling is in practice employed. The resultant cost of reforestation is mostly left to the government, and the costs of this alone seem likely to exceed the royalties of current contracts, ignoring other deleterious effects.

Given the increase of traditional requirements for land and montane forest products, export forestry is unlikely to expand into new areas of forest marginal to the present centres of population in highland Papua New Guinea. (Unfortunately, this concern for the local owners of the forest is not common in most other tropical countries, nor could it be expected in Irian Jaya.) Forestry is therefore likely to develop in sparsely settled montane areas, predominantly the northern and southern slopes of the central cordillera. Some of these very wet, often infertile and exceedingly dissected areas can be expected to resemble the Telefomin situation, and be easily converted to fairly stable non-forest vegetation. With rising fuel (and hence transport) costs, however, very widespread utilisation seems improbable. Montane forest near the coast may be an exception.

More important is the application of the guidelines to non-commercial or small-scale forestry and expansion of agricultural areas. Often unplanned, but nevertheless bound to proceed, they are currently of far greater significance in removing forests permanently.

The historical data show that it is areas of poor soil fertility which are most at risk but that intensive (and forest-fallow) traditional agriculture has achieved stability in some areas. Positive action for reforestation and research efforts at improving garden efficiency (thus reducing dependence on continually opening up gardens in the forest) is needed for the grasslands. Both aspects may imply a need for supplementary nutrients, especially the identification and introduction of suitable fallow species. Papua New Guinea, of course, is already well advanced in this at governmental and traditional levels. The development of *Casuarina* silviculture as early as 1,800 years ago in the Wahgi (Christensen 1975) may be seen as a remarkable achievement, which could well be applied in other tropical countries. Tree planting has been encouraged in government programs and some efforts made to conserve primary forest in threatened areas. More effort is needed to encourage forest regeneration and discourage further clearance on unsuitable terrain, especially along river courses, on steep slopes or at high altitudes. Attempts to shorten tree fallow periods should be regarded with alarm. A forest-agricultural mosaic is the ideal in terms of maintaining the availability of forest products for all, protecting the soil and providing a source of future forest taxa, both plant and animal. Although most of the forest stands would be utilised from time to time, some substantial areas should be retained under forest in perpetuity.

As with White's guidelines for commercial forest, this situation is unlikely to eventuate despite the genuinely favourable attitudes of both governments and people. The individual cost of refraining from cutting timber is too high a price to pay for a generalised benefit which cannot be immediately perceived. The effort to implement the guidelines is worthwhile nevertheless as a reminder for the future (and it incidentally promotes total resource planning). It is vital that small forestry businesses such as local sawmills attempt to manage their timber rights on sound principles to demonstrate that the forest is a renewable resource to local people (and perhaps, through them to large companies as well). The tropical montane forest is adaptable and tenacious as its history shows, but its regeneration requires time which we may think we can no longer afford. We can still less afford to be without the forest.

Acknowledgments

I am grateful to Richard Corlett, John Flenley, Jack Golson, Peter Grubb, Phillip Hughes, Jocelyn Powell and Donald Walker for many ideas about the dynamics of tropical vegetation. Jeannette Hope and Phillip Hughes read the paper and suggested numerous changes, some of which I have adopted.

References

Christensen, O.A. 1975. Hunters and horticulturalists; a preliminary report of the 1972-4 excavations in the Manim Valley, Papua New Guinea. *Mankind* 10: 24-36.

Clarke, W.C. 1976. The maintenance of agriculture and human habitats within the tropical forest ecosystem. Pp. 103-14 in *Ecological Effects of Increasing Human Activities on Tropical and Sub-tropical Forest Ecosystems.* Australian Government Publishing Service, Canberra.

Flenley, J.R., Morley, R.J. and Maloney, B.K. 1976. *Investigation of Quaternary Vegetational History in Southeast Asia.* Report to the National Environment Research Council, London.

Gill, E.D. 1957. The stratigraphical occurrence and palaeoecology of some Australian Tertiary marsupials. *Memoirs National Museum of Victoria* 21: 135-203.

Golson, J. and Hughes, P.J. 1976. The appearance of plant and animal domestication in New Guinea. Pp.88-100 in *Union Internationale des Sciences Préhistoriques et Protohistoriques IXe Congrès Colloque XXII la préhistoire océanienne.* (Prétirage). Centre National de la Recherche Scientifique, Paris.

Hope, G.S. 1976. The vegetational history of Mt Wilhelm, Papua New Guinea. *Journal of Ecology* 64: 627-61.

Kershaw, A.P. 1976. A late Pleistocene and Holocene pollen diagram from Lynch's Crater, north-eastern Queensland, Australia. *New Phytologist* 77: 469-98.

Livingstone, D.A. 1975. Late Quaternary climatic change in Africa. *Annual Review of Ecology and Systematics* 6: 249-80.

Pain, C.F. and Bowler, J.M. 1973. Denudation following the November 1970 earthquake at Madang, Papua New Guinea. *Zeitschrift für Geomorphologie* 18: 92-104.

Powell, J.M. 1970. The impact of man on the vegetation of the Mt Hagen region, New Guinea. Unpublished Ph.D. thesis, The Australian National University, Canberra.

Powell, J.M. *et al.* 1975. *Agricultural Traditions of the Mount Hagen Area.* University of Papua New Guinea, Department of Geography Occasional Paper No. 12.

Robbins, R.G. 1960. The anthropogenic grasslands of Papua and New Guinea. Pp. 313-29 in *Symposium on the Impact of Man on Humid Tropics Vegetation.* UNESCO Science Co-operation Office for South East Asia, Goroka.

Routley, R. and Routley, V. 1977. Destructive forestry in Australia and Melanesia. Pp. 374-97 in J. H. Winslow (ed.), *The Melanesian Environment.* Australian National University Press, Canberra.

Schlichter, H. 1891. Ptolemy's topography of eastern Africa. *Proceedings of the Royal Geographical Society* 13: 513-53.

Singh, G., Kershaw, A.P. and Clark, R.L. (in press). Quaternary vegetation and fire history in Australia. In A.M. Gill, R. H. Groves and I.M. Noble (eds.), *Fire and the Australian Biota*. Australian Academy of Science, Canberra.

Van der Hammen, T. 1974. The Pleistocene changes of vegetation and climate in tropical South America. *Journal of Biogeography* 1: 3-26.

Walker, D. 1970. The changing vegetation of the montane tropics. *Search* 1: 217-21.

Walker, D. and Flenley, J.R. 1979. Late Quaternary vegetational history of the Enga District of upland Papua New Guinea. *Philosophical Transactions of the Royal Society of London* 286B: 265-344.

White, J.P., Crook, K.A.W. and Ruxton, B.P. 1970. Kosipe: a late Pleistocene site in the Papuan Highlands. *Proceedings of the Prehistoric Society* 36: 152-70.

White, K.J. 1975. Effect of natural phenomena in Papua New Guinea forests. (Presidential Address of Papua New Guinea Scientific Society, Port Moresby.)

White, K.J. 1976. Planned development in lowland rainforest in Papua New Guinea with special reference to infrastructure, mining and forestry. Pp. 35-49 in *Ecological Effects of Increasing Human Activities on Tropical and Subtropical Forest Ecosystems*. Australian Government Publishing Service, Canberra.

White, K.J. 1977. Constraints on developing forest industries in Papua New Guinea. Pp. 398-406 in J.H. Winslow (ed.), *The Melanesian Environment*. Australian National University Press, Canberra.

Change in Third World

Chapter 9

At the Tail of the Snake

*William C. Clarke**

> . . . change, as distinct from particular changes, is neither good nor bad in itself, only inevitable. (Spate 1959: 102.)

It was in 1958 that white men first penetrated the forested mountains where live the Bomagai-Angoiang, a small Papua New Guinean community, whose members speak the Maring language. In that year an Australian patrol officer of what was then the Territory of Papua and New Guinea entered the lands of the Bomagai-Angoiang and brought them and their neighbours under control of the Australian Administration. In 1964-5 when carrying out ecological research among the Bomagai-Angoiang, I was interested to know what their lives had been like before that crucial hinge in their history. To some extent a backward view was possible, for the Bomagai-Angoiang lived two days hard walk from the nearest patrol post and airstrip (Fig. 9.1), had no cash crops, and had maintained their houses, dress, intra-community social relations and system of land tenure mostly as these had been before contact with Europeans. They also still possessed their pre-contact knowledge and skills relating to the wild plants and animals in their territory.

Even so, aside from the significant change brought by the much earlier entry of the sweet potato, the Bomagai-Angoiang's

* *W.C. Clarke is Professor in the Department of Geography, Monash University, Melbourne. He was with the Department of Human Geography, Research School of Pacific Studies, The Australian National University, as Research Fellow and Senior Research Fellow for five years beginning in 1969.*

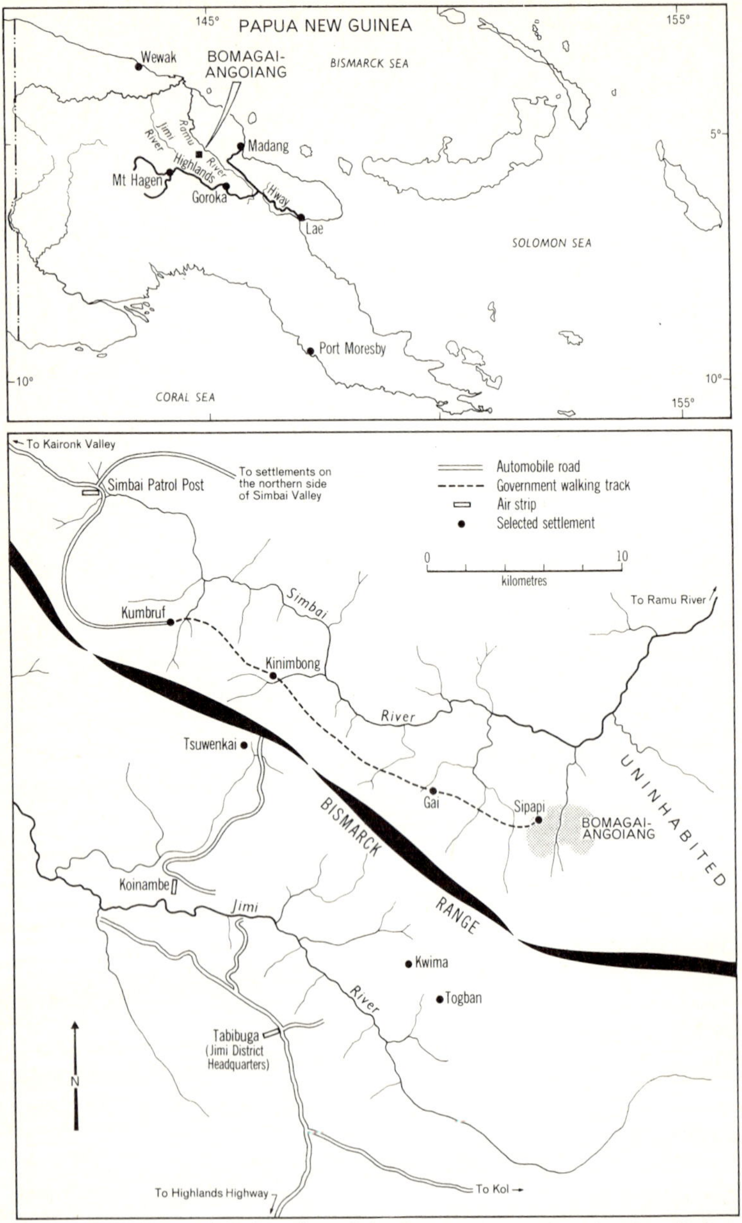

Figure 9.1 The regional setting of the Bomagai-Angoiang in New Guinea.

agricultural system had already been affected for two decades or so by several exotic crops that had passed from group to group along indigenous trade routes before the Europeans arrived in person. Of these recent introductions, the taro-like *Xanthosoma* of tropical American origin was particularly favoured by the Bomagai-Angoiang gardeners as it fitted easily into their tuber-crop complex and gave higher yields than the ancestral *Colocasia* taro, which it closely resembled. Maize, too, was a valued newcomer because it contributed a substantial food earlier in a garden's life than any of the tuber crops. Other innovations that had reached the Bomagai-Angoiang ahead of the Europeans were steel axe blades and bush knives, both of which lessened the labour needed for a variety of tasks and increased the number of trees felled. Of the changes that came during the first few years of administrative control, the most significant was probably the end to the recurrent forays and fights between neighbouring groups (Vayda 1976); this imposed peace brought extended possibilities for travel, changed relations between communities, and a different pattern of mortality.

When I realised that by 1964 European influence had already introduced far-reaching changes among the Bomagai-Angoiang, my attention turned principally to their contemporary situation but I looked back at how they had affected their environment in the past and forward at how their lives and environment might change in the future (Clarke 1971: 184-96). Since then I have been able to make two short return visits, in 1970 and again in 1977. This essay describes some of the changes that have affected the Bomagai-Angoiang between my first and last visits.

Changes in the Population

The basic demographic data collected among the Bomagai-Angoiang in 1965 and 1977 are shown in Figure 9.2; the aggregate numbers are exact to within a few per cent. The forms of the pyramids are roughly accurate, given the difficulty of determining precise ages for most individuals (Clarke 1971: 19). Because of the community's small size, the likelihood of age mis-statements, and what are probably incomplete data on deaths of infants, only some simple observations can be made about demographic change. The average annual rate of population growth was estimated to be 1.6 per cent, a rate that is significantly lower than the estimated present rate of 2.5 to 3 per cent

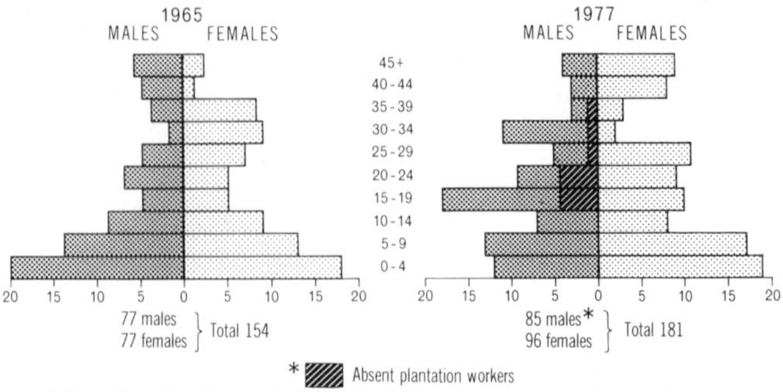

Figure 9.2 Distribution of Bomagai-Angoiang population by age cohorts and sex in 1965 and 1977.

for Papua New Guinea as a whole. The difference is attributable to the Bomagai-Angoiang's comparatively high death rate, which is estimated to be 26 per 1,000 of the population, whereas the national death rate is about 16 per 1,000. The Bomagai-Angoiang's birth rate — estimated by the 'reverse survival technique' — is 42 per 1,000, which is within the range of 38 to 44 estimated for the country as a whole (U.S. Bureau of the Census 1976: 227). As is the case throughout highland Papua New Guinea, pneumonia following an attack of influenza caused the majority of the deaths among the Bomagai-Angoiang. Informants said that around 1972 the community suffered an influenza epidemic which resulted in many of the thirty-two deaths between 1965 and 1977.

During the same period, the number of children under fifteen fell from eighty-three to seventy-six, a decline in the percentage of dependent children in the total population from 54 to 42 per cent — compared with the national average of 45 per cent (U.S. Bureau of the Census 1976: 227). The present low dependency figure for the Bomagai-Angoiang is offset by the absence from the local work-force of eleven men (21 per cent of the men aged fifteen or more) serving as labourers on lowland plantations; in fact, if these eleven are excluded from the total population, the dependency figure rises to 45 per cent. Other changes in the population included the emigration of sixteen girls who went in marriage to other clans and the immigration of fourteen non-Bomagai-Angoiang girls who came in marriage. During the same period there were also five marriages within the community, or clan-cluster, between members of the Bomagai and Angoiang clans. As it was in 1965 (Clarke 1971: 20), so too in 1977 the Bomagai-Angoiang believed it would be good if their population grew; moreover, I could not, in 1977, find evidence that they were aware of the 17.5 per cent increase in their numbers since 1965.

Changes in the Landscape

Walking from the airstrip at Simbai Patrol Post downvalley toward the Bomagai-Angoiang's homeland (Fig. 9.1), the traveller passes through landscapes marked by a gradually increasing proportion of forest over what is surely anthropogenic grassland — a transition that reflects the lessening population density downvalley and perhaps points toward a more recent human occupation of the lower valley. Strengthening the view of diminishing human influence downvalley is the belt of almost uninhabited forest that begins just beyond the Bomagai-Angoiang's territory and stretches southeastward along the northern slopes of the Bismarck Range for some forty kilometres or more toward Bundi, where highland peoples have spilled over mountain passes towards the Ramu River from the densely settled upper Chimbu Valley. In contrast, up the Simbai Valley to the northwest from the lands of the Bomagai-Angoiang, settled clan territories extend over the dissected mountainslopes in a continuous chain to the head of the Valley above the patrol post. To epitomise the Bomagai-Angoiang's location at the edge of settlement, a man from the upper Simbai Valley told me in 1964 that they were a community 'at the tail of the snake', an appellation that implied rusticity as well as marginal location. By 1977 the snake's tail was seen as further drawn out in both respects even though new elements in the Bomagai-Angoiang landscape betokened what might be thought of as economic advance.

In the mid-1960s, the first building to be seen in Bomagai-Angoiang territory by a visitor coming along the government walking track from Simbai was a small, thatch-roofed rest-house on a hilltop known to the patrol officers as Sipapi (Figs. 9.1 and 9.3). Like countless other such edifices then common throughout Papua and New Guinea, this rest-house had been built by the local people to provide temporary shelter for patrol officers, medical teams or any other European visitors. The Bomagai land on which the rest-house stood was designated as the official meeting place between the Administration and the Bomagai-Angoiang together with their immediate neighbours to the west. There, on the surrounding compacted-earth plaza, the local people gathered to listen to messages from the Administration, to be counted, to vote, to hear agricultural officers lecture about plant introductions, to receive medical attention, or to undergo health surveys. Their own houses then lay widely dispersed, singly or in small groups, and mostly on ridge crests that sloped toward the floor of the Ndwimba Basin, the heartland of their territory (Fig. 9.3). From any single place within the basin, only a few houses, at most, could be seen. More commonly, their presence behind sheltering forest or ridge was intimated morning and evening by wispy strands of smoke from household fires.

I do not know whether any patrol officers visiting Sipapi in the 1960s urged the Bomagai-Angoiang to aggregate their houses there, but certainly the people felt the pull in that direction and knew, as Pitt-Rivers (1927: 57) had said four decades earlier, 'European opinion is almost invariably in favour of concentration . . .'. It came, therefore, as no surprise that by 1970, Sipapi had become a place for living as well as for meeting the outside world. Several closely spaced houses, built on roughly excavated terraces, descended from the hilltop rest-house. On some nights almost all the Bomagai-Angoiang men slept here; often they were joined by the women and children, but the women also spent considerable time in 'bush houses' dispersed in a traditional pattern. The most pronounced focus of attraction at Sipapi in 1970 was a Papua New Guinean Lutheran pastor who had come from the Jimi Valley a year or so before. He had persuaded the Bomagai-Angoiang to build a church, which also served as a classroom. This church building, the largest in the settlement, consisted of a grass-thatched roof supported by poles over half walls of plaited grass stems. Within, separated by a central earthen walkway, two rows of poles were laid on the ground — one row for each sex — to provide seats for the Bomagai-Angoiang, who congregated there daily and especially on Sundays to hear the pastor evangelise. A sermon that I attended, which was delivered in Pidgin and translated into Maring by a Bomagai-Angoiang man who had worked on a plantation, had to do with God, the Father who was owed thanks for the beneficence that provided wild game so that the Bomagai-Angoiang need not eat only sweet potatoes and cassava. He also implied that white man had meat even more regularly because of his closer relationship with God.

Food for the pastor and his family was provided by offerings brought from Bomagai-Angoiang gardens as well as from a small garden on ground given to the pastor and his wife by a Bomagai man. The paramedical Aid Post Orderly, a man from the upper Simbai Valley, was said to be similarly provided for by the local people. Because he was absent during my 1970 visit, I could not see what medical supplies were available, but informants said that these sometimes included anti-malarial drugs, cough syrup, antiseptics for sores and cuts, and perhaps penicillin for serious infections. As Bomagai-Angoiang men expressed it, their provision of food and land was a reciprocation to the Aid Post Orderly and the pastor for their services to the community.

When I returned in 1977 commerce had replaced the gospel. By then, the first building inside Bomagai-Angoiang territory was a trade store of the sort common in rural Papua New Guinea: a local-materials shed just large enough for one man to stand in comfortably. Throughout my visit in 1977 the store remained closed because it was too hard work to carry in the supplies to replenish the few shelves. Nonetheless, the area around the store remained a focal point in the community's life, partly because of the symbolism of the store and the track to Simbai and partly because the store's principal owner, who lived in the house adjacent, was also the Angoiang representative in the local government council that had been formed in the lower Simbai Valley in the early 1970s. Sharing his house were two other modern-minded men in their early thirties and next door was the new house of two brisk young men who had recently returned from two years away as plantation labourers. The less traditionally-minded Bomagai-Angoiang viewed this locale as the most exciting part of their domain — almost as a stepping-stone to money and modernisation.

In contradistinction, at Sipapi only a somewhat dilapidated rest-house remained in 1977; the church, the aid post and all the residences were gone. The pastor — who had returned home ill a year or so earlier — had promised to come back, but no one knew when. The church site was now a patch of tobacco plants flourishing in the ash-rich soil. With no apparent sense of remorse, informants said that the Aid Post Orderly had left because the local people had tired of providing him with food and had been stealing food from his own small garden. The houses at Sipapi had been abandoned because of the influenza epidemic — an example of illness-induced fragmentation that is traditional to the Bomagai-Angoiang (Clarke 1971: 100-2) and much of highland Papua New Guinea. However, this dispersal was probably less dramatic than it would have been traditionally; rather than scattering widely over their whole oecumene, most of the Bomagai built houses at or near Mborpe (traditionally Bomagai land), and most of the Angoiang moved to Mangal or Yan'nemfei, both of which are Angoiang lands (Fig. 9. 3). Thus, two mini-villages were formed at Mborpe and Mangal, wherein many more houses were in view of each other than would have been the case in traditional settlements. By 1977, only a few men chose to reside in comparative isolation in houses more than ten minutes' walk from these centres.

House styles and the proportion of houses to persons also changed between 1965 and 1977. The traditional style of Bomagai-Angoiang house (Clarke 1971) was already being replaced in 1970; by 1977, twenty-four of the total of forty-four houses were more of an introduced than of a traditional style. They were larger, higher, squarer, sometimes with an elevated pole floor, had windows, and the walls were of 'blinds' (wickerwork panels of split grass stems). Most of those still built in the traditional low and elongated form also had walls of 'blinds' instead of the traditional materials of *Pandanus* leaves or strips of bark fastened between stakes. The number of houses declined from forty-six in 1965 to forty-four in 1977, giving an average number of occupants per house of 3.3 in the earlier year and 4.1 in 1977. This increase was mainly because of an increase in the number of houses cohabited by husband and wife in contrast with the traditional pattern whereby husbands and wives lived separately.

In the humanised landscape of garden and fallow growth that provides most of the Bomagai-Angoiang's sustenance there were few changes in components; but those components had shifted marvellously — like tesserae in a mosaic — in accord with the principles of shifting cultivation. An old garden in 1965 had become a grove of soil-enriching *Albizia* trees; elsewhere a cool cover of high forest had been replaced by a jungle of tall grasses and vines; or the site of a young garden was now grown with the common secondary tree *Alphitonia* whose light canopy at 10 m shades the weedy *Cyathea* tree ferns as well as remnants of the garden in the form of bananas and papayas still bearing fruit. Among the spontaneous regrowth around sites of abandoned houses were also ghostly reminders of a domesticated past: clusters of the ritually important *Cordyline*, a clump of the banana-like *Heliconia* that provided leaves for wrapping food, or a giant, fountain-like spray of the multipurpose bamboos always planted at settlements.

Within the subsistence gardens the only significant change had been an increase in manioc, or cassava, another recently introduced plant of tropical American origin. The people believed manioc's high-yielding tubers to be excellent for fattening pigs and were also eating more of it themselves, though manioc had certainly not become a major source of food nor was it the meal of last resort as in many overcrowded parts of the tropics. A subsistence innovation in 1970 was a fishpond some 2 m in diameter and 0.5 m deep in whose muddy water there were said to be three fingerlings. The pond had been dug with enthusiasm in flawed imitation of large ponds at Simbai but was too small to be productive and had disappeared by 1977.

By far the most significant change in the agricultural landscape was the spread of what is still the Bomagai-Angoiang's only cash crop, coffee. In 1965 there were no coffee trees; by 1970 a few men had brought in handfuls of coffee seeds obtained from friends or kinsmen in the Jimi Valley where coffee planting and other forms of

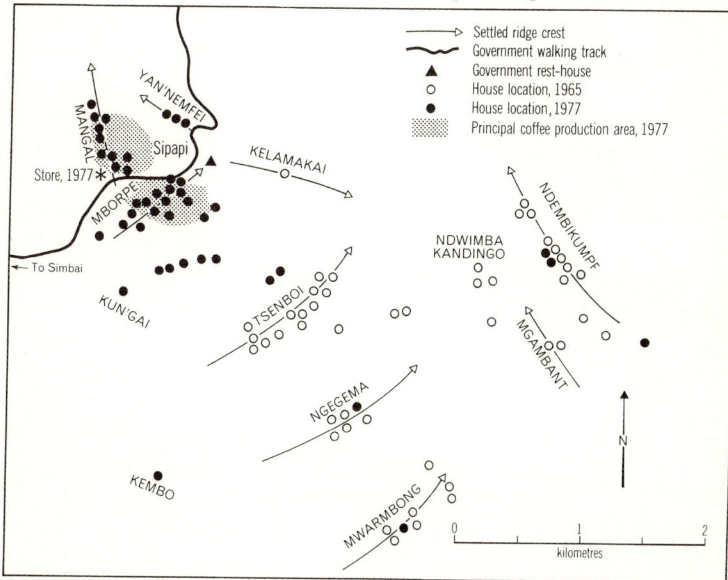

Figure 9.3 Location of Bomagai-Angoiang houses in 1965 and 1977.

development were further advanced. After 1970 (probably in 1973 when the government launched a program to encourage coffee in economically disadvantaged areas) coffee seedlings were also obtained from the agricultural officer at Tabibuga in the Jimi Valley so that by 1977 groves of coffee trees — some already bearing — were well established. Most men had some coffee but separate holdings were still small; the man with the most coffee (planted with the help of his two adult sons) had 933 trees in 1977. The next largest holding was 441 trees, and the smallest was 30 seedlings. Coffee has become concentrated at Mborpe and Mangal on the slopes adjacent to each planter's house although a few men have small groves in the 'bush' for example at Ngegema (Fig. 9.3). The spread of coffee trees removes land from the reservoir of fallowing land available for gardens. In 1977 the man with the largest number of coffee trees, a Bomagai resident at Mborpe, was planting spontaneous coffee seedlings from his Mborpe groves among the subsistence crops of his Kun'gai gardens so that coffee will succeed the gardens. Continued efforts of this sort will push subsistence gardens elsewhere and require either a shortening of fallow or an extension of cultivation into primary forest to the east or up the slopes above the presently cultivated territory — a process, perhaps premonitory, that was going on in one garden at Kembo in 1977. In some ways a dispersion of gardens from the vicinities of Mborpe and Mangal could be beneficial. Especially at Kun'gai there are such large stretches of contiguous gardens that almost all secondary forest has gone. The environmental effects of such concentration, which is very different from the traditional wide scattering of small plots, are uncertain; but what must be the resulting increase in runoff from Kun'gai could have been the cause of a recent, large landslide along the creek that drains the area. Another effect of garden concentration is that the men are probably less mobile within their whole territory, which would mean that less wild game is taken (Clarke 1971: 168). As an example, the four gardens of one Bomagai man in 1965 were scattered at such distances from his Mwarmbong house that the average walking time from house to garden was 35 minutes — time that was contentedly spent passing through secondary vegetation where incidental hunting was often successful (Clarke 1971: 167-8). In 1977 the same man's gardens were all in Kun'gai and Kembo on a straight line path from his Mborpe house at walking times of 4, 8 and

20 minutes, an average of 11 minutes walking through his or someone else's gardens.

A likely further effect of stabilising land under coffee will be a reduction in the planting of subsistence orchards of *Pandanus conoideus,* breadfruit, and *Gnetum gnemon,* which traditionally often succeeded gardens (Clarke 1971: 80-2). As Huber (1977: 9) has described with reference to the Anggor people in the West Sepik Province, so each Bomagai-Angoiang man develops and maintains on old-garden lands 'a portfolio of orchards', the trees of which yield valuable food, especially vegetable protein and oil. A few *Pandanus, Gnetum,* and breadfruit are being incorporated into the new coffee groves, but continued succession of coffee after gardens can only mean an eventual decrease in the orchards as no full-fledged replacements are planted for those mature orchards now sliding into senility.

Whatever the further implications of coffee planting may be, there is no doubt that its introduction, along with even the Bomagai-Angoiang's relatively slow population growth, means an acceleration of what Bennett (1976: 3) calls 'the ecological transition', which he defines as 'the progressive incorporation of Nature into human frames of purpose and action'. Needless to say, that transition is only slightly advanced among the Bomagai-Angoiang but its acceleration, however slow, sets in train many other changes. With regard to the Bomagai-Angoiang's landscape and agricultural system, concern turns to carrying capacity and agricultural intensification. Here these vexed questions cannot be discussed in any detail. No evidence of subsistence intensification was visible in 1977. The Bomagai-Angoiang's agriculture remained the no-tillage, low-labour, high-diversity form of shifting cultivation of 1964-5 (Clarke 1971: ch.3). It seemed that the proportion of grassland to forest or woodland had increased on Kelamakai ridge, and that the walking paths there had also become more deeply incised into the clay subsoil. Both these processes point toward what was probably a concentration of activity on the ridge when nearby Sipapi was the principal settlement; they also suggest the beginnings of a transformation in the direction of landscapes now to be found throughout the more densely settled parts of the highlands. These changes cannot be considered as immediately threatening but they could serve as first indicators that under present technology the territory's carrying capacity has been exceeded and the Bomagai-Angoiang have taken steps along a path leading to land degradation.

Socio-economic Changes

Though spatially isolated, the Bomagai-Angoiang are nonetheless linked to the outside world. Nor are their directly cash-producing activities (coffee and the sale of labour to plantations) the only effects resulting from those links while their subsistence activities remain safely undisturbed on the far side of dualism. Comments have already been made on ways in which even their small-scale coffee production feeds change into their subsistence landscape: outlined here are a few economic and social ramifications that have or may come as a result of their external connections.

An appropriate place to begin is with the generation of new 'needs' that the Bomagai-Angoiang expect to be satisfied by the outside world or, as they epitomise it, the *Gavman* ('Government'). The list is familiar: roads, bridges, schools and opportunities for *bisnis*. The common thread is what Keesing (1978: 181), writing about a remote people in the Solomon Islands, has called 'the seductive powers of money as a lure'. In 1964-5 the Maring already accepted and appreciated money in the form of certain coins but as readily — more so in the case of the women — took payment in basic trade goods. Noticeably by 1970 and more so by 1977 money had leaped upwards in everyone's estimation and had come to have a part in almost all transactions and to be an overt object of desire. A young man who had worked on a plantation and now wanted to migrate to town patiently explained: 'Yes, in my place a man can plant gardens, can have abundant food, can build a house for nothing, but a man *can't get hold of money*'.

What preoccupies people, then, is how to gain cash. In this sense their agriculture has become dichotomous: ecologically, coffee groves are very like *Pandanus* orchards; economically and socially, they are not. Coffee has to do with a spiralling accumulation of money by the individual planter. Or, as one man said, 'I plant coffee thinking of my sons; I want them to be wealthy'. Yet the Bomagai-Angoiang share with many Melanesians what economic planners may see as 'a puzzlingly slow response' to the opportunity to move toward fuller participation in the monetised exchange system of the market (Fisk 1975: 53). Fisk's discussion of such slow responses includes factors that fit the Bomagai-Angoiang well: they live in a state of 'subsistence affluence' and the cash that they gain from market activities, although strongly desired, is to a considerable extent 'needed' only for luxuries and supplements to normal self-subsistent consumption (Fisk 1975: 59-60). The people do require money for taxes and school fees and as a significant part of traditional investments having to do with social relationships, particularly acquiring wives; beyond those requirements, their desire for cash is controlled by their expectations of what is sufficient return for carrying coffee over difficult tracks across a mountain range and a river gorge to Tabibuga. In mid-1977, following the steep decline in coffee prices from their record high, the return was considered insufficient and the Bomagai-Angoiang were storing their harvest (which for the man with the most trees — poorly pruned and too closely spaced — was probably less than 200 kg) in bags tied to their ceilings until the prices rose again. A suggestion that they might not rise was discounted.

The Bomagai-Angoiang's other major way to make money is subject to similarly contradictory interpretations. Certainly, what Keesing (1978: 186) wrote of the Kwaio on Malaita also applies to the Bomagai-Angoiang men who have served one or more terms as plantation labourers.

> Laden with trade goods from company stores, the returning workers enabled their kin to preserve their distant symbiosis with the world economy. The returns for two or three years work were modest but predictable; the employment was certain, and conditions — in terms of housing and rations and workload — were known and adequate. Since plantation work was a way-station on the path to adulthood, almost a rite of passage, the status of copra cutters and 'brushers' was familiar and respected.

So beneficial is the arrangement seen to be by the Bomagai-Angoiang that some surprisingly old and influential men were on plantations in 1977. The other side of the coin is a view such as Meillassoux's (1972: 102) that agriculturally self-sustaining communities 'fulfil functions that capitalism prefers not to assume in the underdeveloped countries: the functions of social security'. That is, the community reproduces the labourers, rears them, makes them available at a low cost for a set period of time, and then reabsorbs them before they can become a social or economic burden on the outside world.

Communities in highland Papua New Guinea were traditionally egalitarian — except for a temporally limited dominance on the part of scattered individuals; but now there is concern that, because of recent changes, certain men or groups are taking permanent control of productive capacity at the expense of others. Restricting consideration to the Bomagai-Angoiang (see Howlett's essay in this volume for a more general discussion), class formation is barely incipient and is unlikely to develop far. It is true that, proportionally, there was in 1977 a great difference in size of coffee holdings from largest to smallest, and a few men had no coffee; but the Bomagai-Angoiang's shared locational disadvantage with regard to access to the factors that make for the emergence of 'big peasants' will make impossible any significant ascendancy by one man. It is easier to foresee problems in the shift from the somewhat general land rights associated with their subsistence gardens (Clarke 1971: 31-2) to the more specific and permanent rights that come with the cash cropping of coffee, but as yet there have been no serious conflicts over land among the Bomagai-Angoiang or between them and their neighbours.

What seems a more definite trend at present is a growing disadvantage to women — almost an intensification of class formation between the sexes. Traditionally, in Maring communities, men controlled the land and did their best to control women's labour, which was essential for production of food as such and as a means of augmenting the pig herd. As Lindenbaum (1976) put it for another part of the highlands: 'a wife is the hand of man'. Following European contact, male dependence on female labour increased even more because of the introduction of coffee as a cash crop and the absence of men away as plantation labourers (Buchbinder and Rappaport 1976). So vital is a woman's labour for picking and for processing coffee cherries to the parchment-bean stage that one Angoiang widower, who had no coffee trees, summarised his situation in a simple equation 'no wife: no coffee'. It is also essential for men, especially mature men, to have women (wives, or sisters if they are married locally) to tend their coffee trees, gardens and pigs while they are away on plantations. Thus, on top of the care of pigs, whose numbers are cyclic, the women are now faced with more coffee trees and fewer men. Intensifying their work is the expectation that the pig herd be increased to its maximum so that there can be an impressive pig-killing ceremony when the men return. For instance, in 1977 my pig census of a sample of Bomagai-Angoiang households (45 per cent of the population) showed a pig/person ratio of 1.02, which is fairly high for the Maring (Rappaport 1967: 93). Considering all aspects of women's functions within the local economy, it seems clear that recently the returns to women have grown more slowly than their contributions. Other aspects of the penetration of the outside world have similarly favoured men: women have gained less access to increased mobility and education; the bush knife has eased their work but not to the extent that the steel axe has lessened men's tasks; and women's access to medical facilities, especially if such are as distant as they are from the Bomagai-Angoiang, is less than men's at the times they may most need them.

Changes in Administrative Concern and the Provision of Services

An analogy can be drawn between the physical landscape and a metaphorical surface of administrative concern. In the one, grassland decreases from Simbai toward Sipapi; in the other, it is administrative concern and provision of services that fade away. In 1958 when the Bomagai-Angoiang and their neighbours were first brought under administrative control, they became the subject of great concern. They were 'newly contacted peoples'; it was adventurous and considered important that patrol officers and other officials proselytise them, explain new ways to them, harangue them on soap and schools and peace and prosperity, count them, examine them, treat them, organise them to build walking tracks, and bring them into the fold. Although first contacted from the Jimi Valley, the Bomagai-Angoiang were soon placed under the jurisdiction of the officer at Simbai in what is now the Madang Province. Given the Bomagai-Angoiang's location, it was not possible for patrol officers to visit very frequently, but through the 1960s and into the 1970s there were official annual patrols as well as patrols for special purposes such as malaria surveys, agricultural development or political education. Europeans from the Anglican Mission at Simbai also visited regularly and provided medical treatment, particularly a clinic for pregnant women and nursing mothers. Not too far away, a primary school was established at Gai under mission auspices. As mentioned previously, the Bomagai-Angoiang became part of a local government council in the early 1970s, and an aid post was established at Sipapi.

The council system remains, but the aid post — which was little valued by the people — has retreated to Kinimbong, a day's walk away. The Anglicans no longer visit. The walking tracks are falling into disrepair compared with their earlier well-maintained state, and the school at Gai has closed. The nearest school now in the Simbai Valley is a government primary school at Kumbruf on the road head from Simbai, but no Bomagai-Angoiang children attend; rather, five boys go to the government primary school at Togban in the Jimi Valley, where the Bomagai-Angoiang have traditional affiliations. Perhaps most symbolic of the withdrawal of the 'zone of concern' towards the government post at Sipapi is that during the first three country-wide elections (1964, 1968 and 1972) special patrols set up a polling station at Sipapi, but during the last (1977) election, two years after Papua New Guinea's independence in 1975, those Bomagai-Angoiang who wanted a vote had to cross a rugged river gorge and walk for four hours to a polling place. I was told that four men had done so. A Bomagai-Angoiang man also told me of the current feeling of his community with regard to the *kiap*, or patrol officer, at Simbai: 'the *kiap* had abandoned us, and we're going to abandon him too'.

At the Tail of the Snake

The process of marginalisation undergone by the Bomagai-Angoiang, and thousands of similarly situated groups across the world, would have been difficult to forestall. Relative deprivation is an inevitable attendant of modernisation and economic development; Papua New Guinea's immature transport network and obstructive topography only intensify the country's developing spatial inequalities. It was inevitable, too, that the concern directed at groups such as Bomagai-Angoiang during their first years of administrative control would fade away as they became remote rather than romantically recent. Today, their economic contribution to the nation is very small, they are getting by satisfactorily on what they have, they do not cause trouble; it is only sensible to direct the limited resources of personnel and money elsewhere.

There have also been structural changes in administration that encouraged a withdrawal of services. The myriad functions performed by a single patrol officer who carried them to the furthest communities of his realm have now been distributed among a number of officers at desks in Simbai, Madang or even Port Moresby. To the Bomagai-Angoiang, whose only experience has been of government coming to them, this lateral and hierarchical distribution of services is still mostly inaccessible and unknowable. Simbai, with its airstrip, several large trade stores, hospital, mission, school and aggregation of government officers, could be viewed as a rural service centre, but even between it and regions like that inhabited by the Bomagai-Angoiang there is what Howlett and Ward (1977: 2) refer to as a 'critical hiatus'. The Bomagai-Angoiang feel the onus lies with the government to come to them; their needs or knowledge are seldom sufficient to drive them to it at Simbai. When they do go, they seem transformed into small, shy 'bush people' who have lost the assertive independence that one finds in them on their home ground. On their part, the recently recruited national officers in the public service are themselves inexperienced and, after the Australian model, are 'desk-bound and procedure-conscious' (Howlett and Ward 1977: 5). Even if they wanted to patrol in the more remote regions, they would find it difficult to get away and unrewarding in terms of promotion. Centres such as Simbai or larger are urban islands in a rural sea, and there is little movement across the boundary between land and water.

It is ironic that the pressures toward marginalisation exist within the framework of a stated national development strategy that strongly favours rural development, the reduction of economic inequalities, and the maintenance or improvement of subsistence agriculture (Central Planning Office 1976). The ongoing political decentralisation of the national government to provincial governments is meant to serve this strategy by bringing decision-making power to local people and making government services accessible to them. Thus, the cause of marginalisation is not at the policy level but lies in the organisational structure of how services are provided and how communication flows within the country. People living in what have come to be remote peripheries find themselves in an ambiguous relationship *vis-à-vis* the more developed centres; they are both tied to them and separated from them. The outside economic system penetrates deeply into their world, yet they cannot grasp it. The relationship is like a bad telephone connection: the link is there, but the intercourse is unsatisfactory to both parties.

Here can be outlined only a few suggestions as to how the connection might be improved and economic development encouraged. It is almost universally agreed that the extension of vehicle roads will better the situation of rural peoples; certainly, the Bomagai-Angoiang are particularly attracted to the Jimi Valley and say that they want to shift from Simbai to the jurisdiction of the Jimi District Headquarters at Tabibuga because they see the road that connects with the Highlands Highway will soon extend across the Jimi River to within a day's walk of their home (Fig. 9.1). However, the improved access that roads bring can also bring a wider diffusion of backwash effects and increased relative deprivation unless there are also changes in the organisational structure that binds together people, policy and economic activities (cf. Logan 1972). Lipton (1977), who argues forcefully that urban bias acts to inhibit rural development, discusses a variety of ways to increase equality and improve rural services. Among many others (Lipton 1977: ch.14) these include: change the reward structure in the public service to encourage rural postings, make those postings for a longer period, and move economists from the central banks to crop-research institutes — in short, 'ruralise' the public service. In a detailed study of information flow and innovation diffusion in the East Sepik Province, Allen (1976: 313-14) makes similar suggestions for Papua New Guinea, urging the appointment of 'liaison officers' to act as links or local-level ombudsmen between the villagers and the bureaucracy. Another approach is to urbanise the countryside, as illustrated in Papua New Guinea by the *maket raun,* which is an attempt to provide rural communities with mobile services on a periodic basis (Ward *et al.* 1978). A number of approaches to rural improvement and more complete information flow between rural villagers, politicians and government officials is discussed in a recent report on development in the Chimbu Province of Papua New Guinea (Howlett *et al.* 1976). The suggested strategies stress an increase in contact or a more sustained contact between villagers and urban-based persons, a need also described by several others with regard to agricultural extension services (Densley and Dick 1978).

To one degree or another these various efforts to push the favoured city into the less favoured country or to give country areas access to the city will act to bring development to remote regions in the sense of increased contact, commerce and communication. Yet I view ambiguously a decrease in the Bomagai-Angoiang's remoteness. They still live in a state of relative autarchy. Economically, they fit perfectly the fifth of Papua New Guinea's eight aims, or national goals (Central Planning Office 1976: 14): 'A more self-reliant economy, less dependent for its needs on imported goods and services and better able to meet the needs of its people through local production'. I think some of the older men share my ambivalence, but most Bomagai-Angoiang would opt — like remote peoples everywhere — for any increase in economic development or contact with the different and seemingly richer world beyond their borders. Because this is so and because trying to translate urban attributes into the country is such a hard task — like walking on the Simbai Valley's steep and slippery tracks — policy should perhaps aim not so much at making the country like the town as making the town like the country.

Acknowledgments

I would like to thank Dr M. Rafiq for advice on demography, Dr E. Ogan for many valuable suggestions, Mr P. Elder for hospitality at Simbai, and A. E. Clarke for counting the coffee trees. I am also grateful to The Australian National University and the University of Papua New Guinea for funding field-work among the Bomagai-Angoiang in 1970 and 1977.

References

Allen, B.J. 1976. Information flow and innovation diffusion in the East Sepik District, Papua New Guinea. Unpublished Ph.D. thesis, The Australian National University, Canberra.

Bennett, J.W. 1976. *The Ecological Transition: Cultural Anthropology and Human Adaptation.* Pergamon Press, New York.

Buchbinder, G. and Rappaport, R. 1976. Fertility and death among the Maring. Pp. 13-15 in P. Brown and G. Buchbinder (eds.), *Man and Woman in the New Guinea Highlands*. American Anthropological Association Special Publication No. 8, Washington, D.C.

Central Planning Office 1976. *The Post-Independence National Development Strategy*. (Papua New Guinea Government White Paper.) Central Planning Office, Waigani, Port Moresby.

Clarke, W.C. 1971. *Place and People: An Ecology of a New Guinean Community*. Australian National University Press, Canberra.

Densley, D.R.J. and Dick, G. 1978. Rural extension and agricultural education. In vol. 1, D.R.J. Densley (ed.), *Agriculture in the Economy: A Series of Review Papers*. Department of Primary Industry, Papua New Guinea.

Fisk, E.K. 1975. The response of nonmonetary production units to contact with the exchange economy. Pp. 53-83 in L.G. Reynolds (ed.), *Agriculture in Development Theory*. Yale University Press, New Haven.

Howlett, D. *et al.* 1976. *Chimbu: Issues in Development*. Development Studies Centre Monograph No. 4, The Australian National University, Canberra.

Howlett, D. and Ward, R.G. 1977. Geographers and development: adaptive strategies or adaptational drift? (Paper read at Fourteenth Annual Conference, Institute of Australian Geographers, University of New England, Armidale, New South Wales.)

Huber, P.B. 1977. *Traditional Land Use in Papua New Guinea: Anthropology and Rural Development*. Discussion Paper 26, Institute of Papua New Guinea Studies, Port Moresby.

Keesing, R. 1978. The Kwaio of Malaita: old values and new discontents. Pp. 180-95 in E.K. Fisk (ed.), *The Adaptation of Traditional Agriculture: Socioeconomic Problems of Urbanization*. Development Studies Centre Monograph No. 11, The Australian National University, Canberra.

Lindenbaum, S. 1976. A wife is the hand of man. Pp. 54-62 in P. Brown and G. Buchbinder (eds.), *Man and Woman in the New Guinea Highlands*. American Anthropological Association Special Publication No. 8, Washington, D.C.

Lipton, M. 1977. *Why Poor People Stay Poor: A Study of Urban Bias in World Development*. Temple Smith, London.

Logan, M. 1972. The spatial system and planning strategies in developing countries. *Geographical Review* 62: 229-44.

Meillassoux, C. 1972. From reproduction to production: a Marxist approach to economic anthropology. *Economy and Society* 1: 93-105.

Pitt-Rivers, G.H.L. 1927. *The Clash of Culture and the Contact of Races*. Routledge, London.

Rappaport, R. 1967. *Pigs for the Ancestors: Ritual in the Ecology of a New Guinea People*. Yale University Press, New Haven.

Spate, O.H.K. 1959. *The Fijian People: Economic Problems and Prospects*. Legislative Council of Fiji Paper No. 13, Government Printer, Suva.

U.S. Bureau of the Census 1976. *World Population: 1975 — Recent Demographic Estimates for Countries and Regions of the World*. U.S. Bureau of the Census, International Statistical Programs Center, Washington.

Vayda, A.P. 1976. War as a process — the Maring case. Pp. 9-42 in A.P. Vayda, *War in Ecological Perspective: Persistence, Change, and Adaptive Processes in Three Oceanian Societies*. Plenum Press, New York.

Ward, R.G. *et al.* 1978. Maket raun: the introduction of periodic markets to Papua New Guinea. Pp. 99-111 in R.H.T. Smith (ed.), *Market-Place Trade — Periodic Markets, Hawkers, and Traders in Africa, Asia, and Latin America*. Centre for Transportation Studies, University of British Columbia, Vancouver.

Chapter 10

When is a Peasant not a Peasant?

Rural Proletarianisation in Papua New Guinea

*Diana Howlett**

> . . . that terrible First Law of Development: to him that hath much, shall much be given. (Spate 1973: vi.)

A significant process of structural change is appearing in Papua New Guinea — the growth of a rural proletariat. The proletarianisation of rural society in the Third World, notably in Latin America, India and parts of Africa, appears without exception to have been associated with colonialism. Sometimes this process was reinforced by pre-existing social structures (Uganda) or religious codes (Hindu and Moslem societies); in other instances a significant rural proletariat emerged only after independence as a result of trends initiated during the colonial era (Latin America). In some countries a rural proletariat was created by specific colonial policies (Kenya), whereas in others such structural change arose from official neglect of certain regions (as in west and northwest Tanzania). Whatever the particular impulse, it would seem that the emergence of rural proletariats in the Third World has been a response to land shortage or landlessness in predominantly agricultural communities, brought about either by increasing pressure of population or by the introduction of capitalist agriculture (used here in the sense of

* *Diana Howlett graduated from The University of Adelaide and then took her doctoral degree in the Department of Geography, Research School of Pacific Studies, The Australian National University, in 1962. After teaching at Sydney University and The State University of New York, she returned to The Australian National University in 1971, and is now Senior Research Fellow in the Development Studies Centre.*

production of agricultural commodities for export), or both. Although Lipton (1977: 112) contends that differentiation in rural society derives from urban rather than capitalist pressures, his argument seems to be less applicable to Papua New Guinea than to countries with either an indigenous urban tradition or a longer period of post-colonial experience.

According to Stavenhagen (1975: 62, 130), six processes of change constitute the essential conditions for the transformation of traditional societies: 'the introduction of a monetary economy, the introduction of private landholding and commercial monoculture, the migration of workers and rural exodus, urbanization, industrialization, and, finally, national integration' with the consequences of 'profound alterations' in the economy, in forms of work and in social organisation.

Some of the pre-conditions which have led to rural proletarianisation elsewhere in the Third World are now evident in Papua New Guinea, and indeed have characterised certain regions for many years. The spatial inequalities in the country's resource and population distribution were largely overcome by less advantaged communities in pre-colonial times by a variety of strategies ranging from warfare to organised trading expeditions. Colonisation and the introduction of a cash economy, although processes by no means as severe as in many parts of the Third World, have nevertheless intensified regional differences. In East Africa, Cliffe (1977: 201) has identified labour-supply areas, cash crop-producing areas, quiescent areas (in which the traditional economy was minimally disturbed), and frontier areas which were opened up by the spread of African population. This schema has general application to Papua New Guinea, although perhaps at present the 'frontier areas' are more apparent around the towns than elsewhere. An inevitable result of inherent and imposed spatial variation is that not all villagers are able to derive a cash income in their clan territories. Some areas are densely populated, some have suffered relative neglect, and in others cash cropping is precluded by environmental or locational constraints. Throughout the country, however, the majority of villagers now have inescapable needs for cash and also a strong desire for it. This is particularly so in the highlands, where the recency of colonialism has left a heritage of two seemingly contradictory motivations for entering the monetary sector. One is the desire to 'catch up' with people in other parts of the country. The other is that cash and the goods it can purchase have been 'traditionalised': cash and business enterprises are widely used to achieve, maintain and reinforce status and prestige according to traditional values. It is doubtful if many Papua New Guineans are yet landless in the sense that they do not have access to land for subsistence. Many, however, are *functionally* landless in that their land is economically disadvantaged by environmental or locational conditions which prevent them from earning a cash income. This is the background against which proletarianisation must be seen.

The inherent spatial differences in the country's resource base have been intensified and modified since colonisation in ways which have resulted in structural differentiation. Social scientists, particularly during the last decade, have documented the country's transformation from the relatively classless society and subsistence-based economy which existed before colonisation to one which is increasingly stratified and monetised. The transformation has at certain times and in particular places been so rapid that social differentiation which elsewhere occurred gradually over many generations has, in Papua New Guinea, taken place within a few decades. The most conspicuous changes, hence those which have received the most attention, have involved the social differentiation of urban populations. These changes are related not only to the process of urbanisation, but to associated factors such as the increase both horizontally and vertically of educational opportunities since the early 1960s; the increasing access of Papua New Guineans to occupations in government, administration and private enterprise; the increased migration of villagers to towns; and changes in political status from dependent colony to independent nation. Indigenous urban society now comprises the élite, the bureaucracy, the petty bourgeoisie, the proletariat, and marginal groups including squatters.

Structural changes and trends in rural society are less readily identifiable. One reason would seem to be the patchy spread of colonial contact and administrative control, which began in some coastal locations in the 1870s and in some interior regions not until the 1960s. Change in rural social structure thus occurred in discrete pockets rather than on a broad scale. Another probable reason, which may be applicable in Papua New Guinea, has been suggested by

Mintz (1974) who commented on the problem of rural proletarian consciousness and the relationship of the rural proletariat to the peasantry, including the 'concealment' of landless workers in peasant communities. Only one direct reference to the existence of landless people is made in the *National Development Strategy* (Central Planning Office 1976: 24-5), and this is in connection with new land legislation intended to prevent, among other things, 'possible conflicts between a landowning class and dispossessed rural people'. However, it is apparent that land-short groups (in the sense already explained) and wage labour for big peasants are more common than is generally recognised. It is likely to be in the interest of richer peasants, many of whom also have local, provincial and even national political influence, to ensure that such concealment is maintained. It is quite possible, therefore, that socio-economic change in Papua New Guinea may be leading to rural proletarianisation, even though this process may have been concealed.

The conventional wisdom of the 1960s in relation to modernisation theory in some respects resembled a cargo cult, in that most exponents firmly believed that modernisation could be achieved if only the vital secret ingredient (the 'big push', 'n-achievement', modernisation crises, balanced growth, unbalanced growth . . .) to enable the transition from less developed states such as peasantries to modern economic systems could be identified and applied (Howlett 1973). However, it has been argued (Howlett 1973) that the condition of peasantry is not always nor even often transitional to some higher and more 'developed' state. Peasantries usually represent a terminal stage of development. Implicit in a condition of 'arrested modernisation' is the *persistence* of the peasantry (see also Mortimer 1975; Low 1976), but this is not to contend that peasant society is therefore static and fossilised, nor to deny the likelihood of differentiation and stratification within the peasantry. Several authors (Finney 1973; Morauta 1974; von Fleckenstein 1974; Gerritsen 1975; McKillop n.d.; Amarshi and Blaxter 1975; Good 1976; Howlett *et al.* 1976) have documented, with varying emphases, the emergence of a class of 'big peasants', a rural élite which has gained control of significant resources and monopolised access to government services and agencies intended for wider dissemination. As Low (1976: 24) has observed, 'Dominant peasants seem to be springing up all over the place'. Inequality in rural areas is now entrenched and seems to be increasing.

Until the last few years of the colonial period, some alternatives to *rural* wage employment were generally available to those lacking cash earning opportunities within their own clans. Thus the over-alienation of land by early colonisers in New Britain, New Ireland and along the north coast of the mainland led to the growth not of an agricultural work-force but rather of an educated and (by now) relatively urbanised sector of the work-force. In later years, and especially during the 1960s, the rapid rate of urban growth attracted a number of migrant job seekers. However, the conditions attaching to rural employment were as important as these alternatives in shaping the overall structure of the paid work-force. Rural employment and wage policies from the end of World War II until the early 1970s were designed to reconcile two basically opposing objectives: on the one hand, to guarantee a supply of labour to the expatriate plantations and, on the other, to protect and preserve village society. Since the regions where labour was required and where it was available were often widely separated, a dependence on migrant workers was inevitable. In the effort to maintain village life, workers could be employed beyond their home province for limited periods only, and repatriation was mandatory for those employed under 'agreement'. The formation of a class of permanent rural wage earners was thereby deliberately inhibited during most of the colonial period. Official policies did not, however, prevent the emergence of a wage-earning rural work-force. Twenty years ago Rowley (1958: 540) observed that 'A class of wage earners, permanently dependent on earnings from wage employment, has begun to coalesce, and will pose significant social, economic and even political problems, as new recruits cut adrift from subsistence agriculture', and he urged the adoption of 'measures necessary to stabilize the conditions affecting those who are or will become permanent wage earners'. Smith (1975: 87) reports that a Board of Inquiry appointed in 1964 to investigate rural wages and related matters (Territory of Papua and New Guinea 1966) also recognised the small but growing number of the rural work-force 'who appreciate the benefits of continuous employment . . . and are content to leave the traditional environment behind them'.

During the 1970s employment opportunities and the rate of new job creation for unskilled rural people in the country's urban centres have declined markedly, although the decision to localise the public service has meant wider employment opportunities for the better educated members of the work-force. Unskilled labour increasingly has to compete with better educated younger people for scarce urban employment. Furthermore, Papua New Guinea's *National Development Strategy* (Central Planning Office 1976: 36) states that

> there will only be a limited number of wage employment opportunities, and no foreseeable expansion of wage employment in urban areas is likely to provide jobs for all seeking work. By 1981 it is likely that 40 per cent of adult males [in urban areas] will be without formal wage employment.

It would appear therefore that villagers from crowded homelands or regions unsuitable for peasant agriculture will in future be obliged to seek employment in the rural sector to a greater extent. Acquisition of smallholder blocks in government-sponsored resettlement schemes is an option available to only a small minority of such people, and it is by no means certain that such land is allocated to those whose need is greatest (Howlett *et al.* 1976: 144 ff.). Cases of 'spontaneous resettlement' of villagers in other rural areas have been documented by Tomasetti (1966) and Ploeg (1975), but this is a precarious strategy which has often resulted in the eviction of the would-be settlers. Although the country has extensive unoccupied tracts, their isolation and terrain are such that they are unlikely to be settled as were the 'frontier areas' in Africa.

It is not known how many Papua New Guineans are in agricultural wage employment, as no statistics are available for those

Table 10. 1. Work-force employed on largeholdings, Papua New Guinea, 1970-1 to 1976-7*

Year ending 30 June	1970-1	1971-2	1972-3	1973-4	1974-5	1975-6	1976-7
Largeholdings	1,255	1,215	1,223	1,172	1,062	950	918
Number employed	51,735	50,629	49,464	52,140	46,933	42,668	41,319

* 'Largeholdings' include government research, experimental and extension stations, and the holdings of religious missions and other bodies, as well as those of private commercial enterprises. Before 1978 they were officially termed 'non-indigenous holdings'.

Source: Bureau of Statistics, Papua New Guinea, *Statistical Bulletins* of 24 March 1977, 24 April 1978, 11 October, 1978.

Parker, M.L. 1973. *Papua New Guinea: An Inter-Industry Study*. Department of Economics, Research School of Pacific Studies, The Australian National University, Canberra.

Peat, Marwick, Mitchell and Co. 1973. *Tourism Development Plan: Papua New Guinea*. Report for Department of Foreign Relations and Trade, Port Moresby.

Robinson, H. 1976. *A Geography of Tourism*. Macdonald and Evans, London.

Samy, J. 1973. Crumbs from the table? The workers' share in tourism. *Pacific Perspective 2: 39-48.*

Simet, J. 1976. From a letter. *Gigibori: A Magazine of Papua New Guinea Cultures* 3(1): 1-2.

Spate, O.H.K. 1965. Progress at Mbananakoro. Pp. 66-87 in O.H.K. Spate, *Let Me Enjoy: Essays, Partly Geographical*. Australian National University Press, Canberra.

Talyaga, K. 1974. *Should We Allow Foreign Research Workers and Tourists in the Enga District*. Institute of Papua New Guinea Studies Discussion Paper 1.

Williamson, P. 1977. Tourism. *Administration for Development* No. 8: 31-43.

Wolfers, E.P. 1971. Tourism. *Newsletter EPW-28*. Institute of Current World Affairs, New York.

Young, E. 1973. Population mobility in the Kainantu area; patterns of movement of the Agarabi/Gadsup people from contact to the present day. Unpublished M.A. thesis, University of Papua New Guinea, Port Moresby.

Chapter 12

The Fijian People

The Spate Report Revisited

*I. Q. Lasaqa**

Since the advent of British rule in Fiji after Cession in 1874, two major views have been expounded about the policy to be followed in the development of the Fijian people and their resources. On the one hand, there was the argument that the Fijians should be ruled by, and be allowed to progress in terms of, Western standards of administration and economic development. They were to be governed also by a system of administration that provided for other communities as well. Such a view tended to emphasise the role of individual effort as the source of Fijian salvation in a highly competitive world, and would appear to have been motivated largely by the interests of private enterprise whose dominant aim was to expand economic activity, thus enabling individual Fijians to become self-reliant, self-seeking individuals cultivating their block of land on a secure title. On the other hand, there was the insistence that the Fijians should be governed in accordance with their ancient customs and traditions. This was the view adopted by Sir Arthur Gordon, the first resident Governor; in due course it became accepted policy. Not only was this dictated by financial necessity, it also made good political and economic sense at the time. Native officials governed their people, with guidance and advice from officers of the central government; in this way the Fijians controlled their own local affairs using tradi-

* *Isireli Lasaqa is Secretary to the Cabinet of the Government of Fiji. After graduating from The University of Auckland, New Zealand, he took his doctoral degree in the Department of Human Geography, Research School of Pacific Studies, The Australian National University, in 1969.*

tional institutions. This approach ensured that the Fijians would have abiding faith and trust in the government which they believed would do what was good and proper for Fijian interests.

Despite some uncertainties in the 1920s and 1930s when Fijians were ruled directly by expatriate Provincial Commissioners appointed by the central government, the Fijians remained loyal to their chiefs, and their social and economic life continued to be based on their communal organisation. At the same time, however, there was an increasing awareness of the need to devise a system of administration that would square with the needs of the time.

By World War II, there was a better understanding of the principles of native administration which Gordon had defended some sixty years before. British experience in colonial Africa had shown these to be practical and acceptable. The proposed system was to be based on the principles 'of building on the existing organisation and ideas of the people, of leaving it to the people themselves, assisted by sympathetic advice, to devise and develop their own institutions' (Sukuna 1944a: 1). Gordon also argued that 'the only true political development proceeds from within by evolution'. Traditional Fijian social and political institutions were still intact to serve as the basis of post-war development. The provincial and *tikina* (district) boundaries that were adopted corresponded closely to the traditional political units, and local consensus played a major role in the selection of representatives to *tikina* and provincial councils. Inevitably the traditional leaders in society were to become leaders in the new administration. These principles provided the foundation on which the Fijian Administration was launched on 1 January 1945.

The Purpose of the Fijian Administration

When introducing the Fijian Affairs Bill in the Legislative Council on 24 February 1944, Ratu Sir Lala Sukuna (1944b: 36) said that its purpose was 'to train chiefs and people in orderly, sound, and progressive local government better to fit them eventually for the give and take of democratic institutions'. But there was a fundamental conflict which was probably unavoidable. While modern democratic institutions demanded the full participation of the people and the acceptance of their wishes, Fijian society was based on hierarchy and ascribed status. How to change from one to the other was not only problematical, but its timing was even more difficult to determine accurately.

However, during the following decade the Fijian Administration concentrated its efforts on attempts to improve conditions in the villages and on the better use of land by Fijians themselves. To this end village amalgamation was encouraged so that in each case a large pool of able-bodied men would be available for village clearing and house maintenance. A scheme of compulsory saving based on copra income was introduced, largely to help villagers to build their homes in permanent materials, thus freeing 'men from the incessant house-building that goes on everywhere' (Sukuna 1944b: 36). As for the better use of land, the village was to be the base because it does not disrupt the community, preserves the communal spirit and avoids social friction. Thus, the *galala,* men exempted from village social services, were not going to receive active and official encouragement. Any land development was to be based on the village program of work, devised by the *turaga ni koro* (village headman) and confirmed by the *tikina* council. Thus Fijian development during this decade was taking place within the framework of the communal system. The Administration accepted the policy that Fijian economic development should be based on local social institutions, many of which were given legal status through the regulations made by the Fijian Affairs Board.

The Problem

By the early 1950s it was becoming clear that various social, economic and political factors were having a significant impact on Fijian society. Rapid population growth, especially of the Indians, was creating an increased demand for idle Fijian land; at the same time the Indian community was establishing a greater control on the economy of the Colony. More and more Fijians were passing through the education system and, with widened horizons, young people were looking for better opportunities outside the village system. War service also gave the Fijian a new view of life and the world beyond. New associations were being formed; in some instances Fijians were beginning to question aspects of their traditional society, including the communal approach to development. Moreover, concern for Fijian welfare was to become a major feature of government policy in the late 1950s and beyond. This concern was expressed by the Secretary of State for the Colonies (1952):

> I should be grateful if you would give consideration to the

> problem of assisting the Fijian people to play a greater part in the economic life of the territory. I fully appreciate that the traditional Fijian way of life, and particularly the communal approach to economic affairs, limit the opportunites for individual enterprise and reduce the incentive to the individual to improve his economic position. It is debatable, therefore, whether any rapid change in the tempo of economic development of the Fijians can be brought about without some sacrifice of the traditional customs and way of life, the value of which I fully appreciate . . . The problem is, therefore, to decide on the ways and means and the tempo at which development can take place . . . I consider that the time has come for it [the problem] to be seriously re-examined by the local authorities.

In his reply, the Governor of Fiji (1953) recognised the weakness of the communal system as a means of Fijian economic development; he also suggested that it could not be abandoned overnight, but that the Administration should encourage individualism and the co-operative movement. He outlined plans for increased production of cocoa and bananas, and a scheme of provincial development to be funded by rents from leased Crown Schedule A and B lands, and he also hinted at the need to give enterprising Fijians a secure title over the land they cultivate. Clearly the Colonial Administration was in a dilemma. The official interpretation of the Deed of Cession guaranteed the protection of Fijian interests, and official policy favoured gradual development of economic activity based on the traditional institutions. Individual enterprise was tolerated though not actively encouraged. This policy ignored the increasing Indian population which was competing and outstripping the Fijians in most spheres of economic activity. If production were not boosted quickly, the Fijians might be relegated to a subordinate position in their own country: the situation thus had grave political implications.

Because of the complexity of the problem of Fijian development, the central government was convinced of the need for a competent and independent assessment of the situation. O.H.K. Spate of The Australian National University was invited to undertake this assignment. The choice was particularly apposite: Spate had been associated with T.W. Swan and C.S. Belshaw of the same University in an assessment of the problems of economic development of Papua and New Guinea in 1952-3, and their report became 'a basic document in any study of the post-war economic policy in Papua and New Guinea' (Hasluck 1976: 140-1). Furthermore, his experience in India was to prove invaluable in seeing Fijian problems in a much wider multi-racial context.

The Spate Report

Professor Spate was asked to do two things (Spate 1959: 1):

I. To assemble and assess such data as can be obtained on the economic activity of Fijian producers, with special attention to the effects of their social organisation on that activity;

II. To consider how far the Fijians' social organisation may be a limiting factor in their economic activity, and to suggest in what ways changes in that organisation might be desirable.

From the very beginning, Spate (1959: 1) recognised the complexity of the subject, for it demanded an analysis not only of economic factors *per se,* such as the problems of capital and techniques, markets and organisation, but also the social and psychological factors, and the role of the Fijian Administration.

The society

Spate noted that the Fijians owned 83 per cent of all the land, the bulk of which is unsuited to commercial agriculture. The Indian population was rapidly increasing, and dominated sugar cane production, the most rewarding crop, and the service industries. As against this, the Fijians controlled half the production of copra and all the output of bananas (although this is no longer an export crop). In economic terms the Fijians were and remain the weaker of these two communities. Inter-racial relations were remarkably good but the equilibrium was unstable (Spate 1959: 5). It appears that the position has not changed markedly during the last twenty years.

The basic approach to Fijian economic development fostered by government was through the Fijian Administration and the communal system. It was noted that these institutions, designed for non-economic ends, were largely incompatible with modern approaches to development which demand professional expertise in, among other things, capital management and market arrangements. Spate (1959:6) noted that '*vaka Viti* [the Fijian way] was designed, and beautifully designed, to do quite other things than those now demanded of the Fijians . . .'. While accepting 'common descent, common faith, and common interest' as the basis of Fijian society, Spate noted also that cohesion was a great strength, especially in the old days. In the new situation the society and its leaders were not always easily able to adjust to modern demands. Thus consanguinity and status, as bases of social relations, were not adaptable to the demands of contract and individual freedom which money-making urgently demanded. Furthermore, there was a fair degree of particularism in each social unit when land matters were considered. These factors still apply today, although in some situations they are probably blurred by the demands for Fijian solidarity at the national level in the face of competition from other communities, especially the Indians.

Spate (1959: 8) noted also that there was a certain degree of ambivalence in Fijian society: although the goods of Western civilisation were desired, there was always a harking back to the old ways. The Fijians had to come to terms with modern economic life and its rules; they could not enter the cash economy solely on their own terms; there would have to be some degree of give and take and the society would have to adjust to the needs of money and contract.

The land problem

After considering the land tenure situation Spate concluded that the codification of ownership by the *mataqali* (the recognised basic land-owning unit) froze a situation which was essentially flexible. The land use pattern confirms this flexibility even today, especially when subsistence cultivation is involved. At the same time *mataqali* ownership had led to considerable disparity in the distribution of land which even *yavusa* (the widest kinship group) ownership would not solve. Spate considered, however, that secure individual title on a leasehold basis provided the long-term solution to unequal distribution of land among Fijians.

The Reserves policy, whereby *mataqali* set aside a proportion of their land for their own use leaving the balance available to others (including non-Fijians for leasing), had been criticised as being over-protective. Spate's (1959: 21) view that 'Reserves might become merely static defences for an inert garrison' has some validity considering those reserves in the cane areas which have fallen into disuse. Fijian land-owners do not seem to appreciate their moral responsibility to ensure that good land must be put to effective use, not only in their own interest but also in the interest of the country as a whole.

But the Reserves policy was formulated largely to safeguard the interests of Fijians, especially in the light of the increasing Indian population in the 1930s. Such a policy was necessary if Fijians were to agree readily to the leasing of some of their land to others: they needed assurance that, as land-owners, sufficient land had been set aside for their exclusive use.

The burden of obligations

Although Spate agreed that the world would be a dull one if stripped of some of the extravagance associated with celebrations following births and marriages, he believed (1959: 26) that the economic effects of such expenditure are clearly disastrous and become even more so when the means of livelihood are used in this way, such as the slaughtering of a working bullock for a wedding feast. But there is wide variation in the burden created by social obligation, and this variation applies also to the ability of the individuals to meet their social responsibilities. Some of the burdens are part of the traditional subsistence pattern of village life, like house building and maintenance, and village clearing, all of which were codified in the Fijian Regulations. Some matters, like *kerekere* (the practice of borrowing amongst kin at the will of the borrower), were so small and petty that one wonders why Regulations were necessary to limit their incidence. Like Spate, it is difficult to believe that people must be told not to starve themselves or leave their houses in disrepair when land is available for planting and housing materials are in abundance. But Spate concluded (1959: 25) that most of these obligations are matters which could be resolved only by the Fijians.

House building and maintenance were a particular burden in Fijian village life, the improvement of housing being a central theme in Sukuna's post-war development policy. By building permanent houses in the village, it was hoped that more time would be spent on the production of cash crops. But Spate noted (1959: 27-8) that what Sukuna had envisaged as a means to an end had become an end in itself, and that time saved through permanent housing was not always spent on cash-generating activities. He perceived the importance of a regular source of income and a proper forward-looking plan to be essential elements in an effective housing program. Spate also advocated the idea of a building team in the rural areas, the need for some professionalism in village housing and the erection of a couple of traditional houses on the *rara* (village green), one for important guests and the other as a meeting house and centre of community life.

Fijian Administration and Development

Spate rightly concluded (1959: 39) that the Fijian Administration was the organised political expression of the communal system and thus its influence pervaded all aspects of Fijian life. Moreover the village work program governed much of the Fijians' activities. Although it had been historically necessary to set up a separate Fijian Administration, it was thought that it tended to emphasise a separatist outlook among Fijians and their officials. The principle of gradual change in the Fijian Administration was accepted, with a view to preparing the people for 'give and take of modern democratic institutions', but change in the administration would have to square with the rapid shifts occurring in Fijian ways of life. Given the lack of clear leadership, especially at the local level, there was no alternative to the existing system.

Fijian institutions, their mode of operation, and their aims were determined from the top and at the centre; there was little delegation of power on local matters down to the provincial and *tikina* levels for about twenty years after 1945. Because of the lack of suitable staff and experience at the grassroots, the Fijian Affairs Board was naturally reluctant to delegate decision making on local matters, but Spate pointed out (1959: 36-7) that experience would never be gained without such delegation. Yet there was 'no disposition to embark on the uncharted seas of local government'.

The appointment of Economic Development Officers to organise cash cropping was supported by Spate, but he thought their work should be based on the village program of work and collective effort. This appears to have ignored the fact that economic crops were cultivated on an individual basis on *mataqali* land. This placed these officers in a difficult position because the collectivist approach was applicable mainly to social activities and communal facilities such as the construction of a church or school.

The co-operative movement had undoubtedly played a significant role in promoting 'self-help, self-management, self-discipline' among Fijians: it helped in the handling of copra and in the distribution of consumer items. But at the time of Spate's investigation the Fijian Administration seems to have given only token recognition to the value of the movement. Among the officials of the Fijian Administration and those of the Co-operative Department, its value probably lay in the view that it was in accordance with the traditional communal system. This view missed the point that co-operative societies had economic motives and their methods possessed a rationale quite separate from that of the Fijian social system.

Spate supported, with some qualification, the idea of a tax on certain produce, rent and royalties recommended by McDougall (1957) as a means of promoting economic development. He noted various disincentives about the proposal and the suggestion that the Fijian Affairs Board might retain 20 per cent of the amount to be used, at its discretion, on specific projects. Proposed also was the idea of inter-province lending, which Sukuna suggested in 1947 (Sukuna 1948: 11). Spate noted also the idea of a land tax which Belshaw had tentatively suggested as a means of financing provinces but he doubted if it could be implemented rapidly because it was a major departure from current practice and would not be easily understood by the general public.

When discussing the role of the communal approach to Fijian development, Spate reviewed (among others) the progress made in Daku, in the Rewa delta; the bus service at Cautata, Tailevu; and the housing scheme at Naqali in Naitisiri based on banana sales, rice and cane production in parts of the Sigatoka valley. He noted that in all instances authoritarian leadership and in some cases outside official support were major factors. There was local patriotism, but the question of continuity in leadership remained unresolved in most cases. Most of the projects based on the communal approach did not last for more than a few years because they lacked leadership and business management. Spate thus concluded that economic development must, in most cases, be based on individual effort and the appropriate know-how, with the necessary support given by the officials of the Fijian Administration working as a team with the technical officers of the central government.

The basic recommendations

Spate (1959: 9) set out his position as follows:

> for the Fijian countryside the objective should be a community of independent farmers living or working on holdings heritable, and alienable at least between Fijians, but retaining in each village or Old Tikina area a common centre — church, school, guesthouse, parish hall, chiefly residence — where the old dignity which the *koro* is so rapidly losing might be recaptured, and relying for essential joint services on some equivalent of a Rural District Council; a community less the old frustrations (though doubtless with its own), but not less rich in real satisfactions.

This was the objective to be accepted and to be attained slowly by evolutionary means. There was already a tendency towards it throughout the Fijian countryside but its development would have to be encouraged and guided with pace and strategy suited to the needs of each district. This view did not imply the disintegration of Fijian society as some believed. The individual farmer would not be far from his village and would remain within his old *tikina* boundary. Even if he moved to another *tikina*, he would still be among his own people 'sharing a common cultural tradition' (Spate 1959: 96). In this way the individual Fijian would be placed in a position to face and determine his own destiny: he would work out his own work program — it is to be hoped in a more realistic way.

Consonant with this recommendation was the suggestion to give *tikina* and provincial councils a measure of autonomy in local affairs. The provincial councils especially must be given executive power to determine financial and other local matters dealing with development. At the same time the Fijian Regulations, set as they were in traditional social custom and responsibility, needed a complete overhaul because they were out of tune with individual aspirations. With the encouragement of individualism, the official program of work would cease to form the basis of Fijian economic development in the rural areas. The need for professionalism in Fijian economic development was emphasised as well as the close co-operation of officials of the Fijian Administration with those of central government.

Spate suggested that a larger social unit than the *mataqali*, either the *yavusa* or the *koro*, might become the land-owning unit in order to help even out the disparity in land holding. A land tax as a means of financing the province needed much closer examination. A more definite development structure, from the village through the *tikina* to the province, might ensure the better assessment and co-ordination of development work by the economic development officers. Non-agricultural activities, such as fishing, transport operation, mining and forestry, appeared to need greater encouragement and more expert advice.

Finally, the communal approach, as demonstrated by various projects based on it, could not be regarded as the only mode of operation to be adopted throughout the country as the vehicle for Fijian economic development.

A General Evaluation: The Last Twenty Years

There can be no question about the influence of the views and recommendations made by Spate. In many instances the influence has been direct, with the Fijian Administration and the central government adopting a suggested remedy. In others acceptance has been slow and only after a roundabout course and the passage of time have conclusions been reached similar to those of Spate twenty years ago.

Soon after Spate submitted his report, the colonial government appointed the Burns Commission (1960) to review Fiji's natural resources and population trends. That body concluded that the Fijian Administration should be abolished; it was a 'government within a government' and a serious handicap to Fijian advancement and the general progress of the Colony. This view did not gain support from the Fijians and its fate was understandable and predictable.

However, the opinion that the Fijian Administration needed to be reviewed in the light of new conditions found much wider favour. The changes that have occurred in Fijian society and, particularly, the loosening of the social order clearly required a closer look at the basis of Fijian institutions and local councils. This led to the formation of a Committee of the Council of Chiefs in 1962 to review the role and purpose of the provincial councils. This Committee recommended that greater responsibility be given to the provincial councils in local matters and that they be made more directly responsible to the people through a system of direct election.

Provincial councils should also be responsible for appointing their own staff except the *Roko* (the Fijian official in charge of a Province), formulating their own budget and making by-laws along the lines of rural local authorities. The Committee's proposals needed to be worked out in detail and explained to the Fijian people. This was undertaken by the late Dr Rusiate Nayacakalou who in 1964 managed to obtain a wide measure of support for the proposals. All this led to the reorganisation of the provincial councils during 1965-6 and the first election of members of such councils in 1967. These changes were foreseen by Spate who felt that local level institutions would never learn to shoulder responsibility if they were not given the opportunity to make decisions on local matters. Such institutions also needed to be based on democratic principles of local government.

On the financing of provincial councils Nayacakalou (1964) suggested that a fresh look might be given to the question of the land rate. An exercise was undertaken for Tailevu Province which showed that a rate of about 2 per cent of the unimproved capital value of all Fijian land in the Province would be cheaper for 60 per cent of all land-owning units and the Province would improve its revenue by 66 per cent. In this way the Tailevu Provincial Council would be able to raise adequate revenue to meet its administrative expenses. The land rate was adopted in Tailevu in 1968 followed by Kadavu, Ba and Rewa during the next two years, but has not been further extended. The administrative difficulties with which the system was fraught, and which Spate warned about in 1959, have not been completely overcome.

The Fijian land tenure system has remained unchanged; it is likely to remain unaltered in perpetuity, for shifting the basis of land-ownership from the *mataqali* to a larger or smaller social unit is likely to create instability. The *mataqali*, as the unit of ownership, has been set for nearly a century, and it is difficult to believe that Fijians will adapt to another unit easily without upsetting the existing social relations. However, social change has influenced the society deeply and Fijians themselves already want to lease land and obtain better security to enable them to get on in life. This can be seen as part of the Fijians' attempt to embark on the development of their land in a professional way: they are moving from the communal system and the village program of work as the basis for their advancement.

Although certain communal obligations have ceased to be

regulated by law, others continue to be performed, including ceremonials associated with deaths and births. Admittedly their incidence varies: some can be crippling in their economic and financial effects. Yet greater economic development during the last twenty years has enabled the Fijian to continue to afford such ceremonials and it is unlikely that they will be abandoned since this would strike at the very root of Fijian society and the Fijian identity.

One of the major changes that Spate recommended was the overhauling of the Fijian Regulations which governed Fijian village life in minute detail. These were abolished and villagers were given the opportunity to determine their own activities. The decision to abandon the village work program, which determined the use of land and time allocated by the inhabitants, was in keeping with the growing sense of individualism in the countryside and what was already the rule in the urban scene, where wage and contract largely determined individual action.

Spate had advocated that the future for the Fijians lay in the evolution of a community of independent farmers throughout the countryside, but this was not readily accepted because it was seen as a process of breaking up Fijian village society. However, government policy in the late 1960s led to a program of land subdivision and development in areas still vacant and unused. Set schemes were formulated in places like Lomaivuna and elsewhere on Viti Levu, and on Vanua Levu and Taveuni Island. This program was supported by a coconut subsidy scheme which encouraged Fijian land-owners to subdivide their land for new coconut plantings, not only on the two main islands but also in the maritime provinces. No doubt greater land development has occurred in the last twenty years, especially on an individual basis. There has been increasing awareness of market and economic forces by the Fijian in the rural areas, and a growing appreciation of options and choices open to him in the development of his resources. As a result of greater awareness and sophistication he is also better able to assess opportunities for increased participation in the rural cash economy. All this has enabled villagers to place their development efforts, through the advice of the technical officers of the central government, on a more professional footing than was possible through the village work program.

Conclusion

The Spate report provided timely stimulus to Fijian progress during the last twenty years. Spate's assessment of Fijian life, set in the context of traditional society and administrative structure, was clear-cut. The report was forward-looking for it tried to suggest the lines of progress that Fijian society might pursue in order to remain viable. It tried to outline some of the ways in which the society might adjust in order to come to terms with the modern world of contract and commerce, so that Fijian social and economic interaction would not be entirely dependent on traditional mores.

The Spate report was a comprehensive and incisive analysis of a complex situation. For the first time in the history of modern Fiji, the Spate report offered an authoritative analysis of the Fijian situation by a highly trained mind, at once sympathetic and objective. It questioned certain approaches to Fijian development based on the Fijian Regulations and operated through the Fijian Administration. It offered a new direction, much of it evolutionary in nature, towards which Fijian society might develop in its aim to modernise. On the whole, changes in the Fijian Administration and Fijian life during the last twenty years have kept fairly closely to the major signposts set up by Spate.

Spate stressed that in its quest for modernity, Fijian society must reduce its emphasis on status and tradition, and be more concerned about the fitness of individuals for particular functions. Fijian society accepted this challenge and responded positively to this prescription so that today Fijian institutions, both local and national, aim to seek out and recruit Fijians on merit, not status. The fact that Fijian traditional leaders themselves have been in the vanguard of this attitudinal change is a clear indication that they and their people are ready and willing to make the adjustment necessary for social viability and economic vitality.

Although Fijian society continues to face particular problems, of both an administrative and social nature, the analysis of the past has enabled the people to face the future with courage and confidence. A new leadership, with discipline, confidence and imagination, is abroad. These are vital factors for the future progress of the Fijian community because they are qualities it will need in large measure if it is to hold its own in a multi-racial Fiji. It is always a terrible responsibility to be asked to advise on the path to be followed by a whole people and their society. I am sure Spate realised this fully and he can have the satisfaction of knowing that his Report has had considerable influence for the good of Fijian development during the last two decades. At the same time his Report has provided a base from which the Fijian people can confidently face the challenges of the future.

References

Burns, A., Watson, T.Y. and Peacock, A.T. 1960. *Report of the Commission of Enquiry into the Natural Resources and Population Trends of the Colony of Fiji 1959*. Legislative Council of Fiji Paper No. 1, Government Printer, Suva.

Governor of Fiji to Secretary of State for the Colonies 1953. Despatch No. 420, 19 August 1953.

Hasluck, P. 1976. *A Time for Building: Australian Administration in Papua and New Guinea 1951-63*. Melbourne University Press, Melbourne.

McDougall, R.S. 1957. *Fijian Administration Finances*. Legislative Council of Fiji Paper No. 35, Government Printer, Suva.

Nayacakalou, R.R. 1964. Report of the Rural Planning Officer. (Unpublished typescript.)

Secretary of State for the Colonies to the Governor of Fiji 1952. Despatch No. 460, 31 October 1952.

Spate, O.H.K. 1959. *The Fijian People: Economic Problems and Prospects*. Legislative Council of Fiji Paper No. 13, Government Printer, Suva.

Sukuna, J.L.V. 1944a. *Fijian Affairs. Reconstruction Paper, No. 2*. Legislative Council of Fiji Paper No. 22, Government Printer, Suva.

Sukuna, J.L.V. 1944b. *Fiji Legislative Council Debates* (24 February 1944).

Sukuna, J.L.V. 1948. *Secretary for Fijian Affairs (Annual Report, 1947)*. Legislative Council of Fiji Paper No. 52, Government Printer, Suva.

Chapter 13

For Whom Do the Planners Plan?

An Inquiry Into the Status of City Planning in Indonesia

*W. Donald McTaggart**

'Cities are formed through the geographic concentration of a social surplus product, which the mode of economic integration must therefore be capable of producing and concentrating' (Harvey 1973: 216). But if the existence of cities is a symptom of the ability of a social system to effect this level of production and concentration, it says nothing about the particular social mechanisms involved. Cities have come into being, flourished and declined under many different social systems. This assertion contradicts Wirth's (1938) central thesis that such factors as density of population, size of agglomeration, level of interaction and high pace of activity are universals, and are primary forces in shaping the lives of urban residents. Wirth wrote within the mainstream of orthodox Western thinking about city development. Cities in Western Europe did play a crucial role in breaking the grip of medieval feudalism (Weber 1958); they were centres of rationality and modernisation. They also played a crucial role in the emergence of industrial capitalism in the nineteenth century (Weber 1899), and in most Western countries they soon contained the bulk of the population. It is no surprise that such a positive view of the role of the city in Western economic and

* *W. Donald McTaggart is Professor of Geography at Arizona State University. He took his doctorate in the Department of Geography, The Australian National University, in 1963. He then taught at The University of Malaya in Kuala Lumpur before moving to Arizona in 1971. In recent years most of his research interests have focused on Indonesia.*

social development was soon paralleled by advocacy of a similar view for less developed societies.

Orthodox analysis of the urban situation in the Third World has tended to suggest that urban growth, despite its attendant problems, is inevitable, and that planning should aim to minimise its adverse effects. Nonetheless, planners and politicians, recognising the cultural superiority of the city over the countryside — the Great Tradition as opposed to the Little Tradition (Redfield 1953, 1956), should accept the need for a period of painful adjustment, a period of urban poverty, misery and even unrest. It has been argued (e.g. Breese 1966) that to suppress urban growth would be to suppress the process of development itself.

Less orthodox writers have become much more sceptical about the benefits of urbanisation in underdeveloped areas. Urbanisation, it is suggested, does not necessarily entail a transformation or 'modernisation' of existing economic habits. Many of the migrants who flock to the cities continue to subsist within an economic system or sub-system which is, in reality, still a peasant system. The 'modern' sector is difficult to enter, and remains largely the preserve of large-scale and international enterprise, for which the emergence of urban centres in Third World countries is an essential condition for profitable activity (Santos 1971; McGee 1974, 1976; Gomhan *et al.* 1976).

Urban growth has been proceeding at a rapid pace in most countries of Southeast Asia, including Indonesia. But while there is some correlation between levels of urbanisation and various indices of economic development in the region as a whole, there is no consensus about the kind of national urban policy that should be adopted in the countries concerned. Singapore is a committed city-state, Malaysia is actively developing major urban regions, Indonesia tolerates urban growth, and the Khmer Rouge government of Kampuchea appeared hostile to the very idea of the city (Allen 1975). It seems that attitudes to urban development reflect the broad ideological postures of the respective governments. Those whose developmental strategies are posited on acceptance of capitalist principles tend to favour urban growth as a dynamic and active element; those with socialist views take an opposite stance.

Indonesia appears at first sight to hold an intermediate position. In 1971 its population was just over 119 million and was expected to reach almost 142 million by 1978 (Hardjono 1977: 1). At the former date about 17 per cent of Indonesia's population were considered to be urban — a relatively low figure. But rapid urban expansion is not considered to be a feasible solution to Indonesia's population problems; redistribution to unoccupied rural areas and a vigorous family planning program are advocated instead. It is hoped in this way to relieve the pressures on the cities. It is not immediately clear whether Indonesia is aligning itself with Western capitalist notions of development and their implications for the dynamic role of the city, or with socialist emphasis on development through a transformation of the countryside. Closer analysis, however, shows that in terms of the management and planning of its urban space, Indonesia is utilising concepts compatible with capitalist rather than socialist forms of development, and that there is congruence between overall economic policy on the one hand and city development on the other.

Currently Indonesia appears to be adopting urban planning policy unconsciously, or at least subconsciously. A mode of city planning is being utilised ostensibly because it is 'modern' and is therefore consistent with the government's desire to promote 'modernisation'. In fact this aspect of modernisation unwittingly involves the administration in a struggle to uproot many traditional aspects of Indonesian life, and to replace them with forms dominated by considerations of commercial profit. This process is documented here.

Economic and Social Policy in Indonesia

Two periods can be distinguished in Indonesian economic development since independence. From 1950 until 1965 Indonesian politics were dominated by President Sukarno, who followed policies generally regarded as strongly nationalistic, with a powerful appeal to the people to unite to throw off the yoke of colonialism. Foreign capital, which had formerly dominated the export sector of the economy (rubber, tin, palm oil, tobacco, petroleum products and tropical timber), was no longer welcome, and many enterprises were nationalised. The decline of economic performance during this period was exacerbated by external political entanglements, such as confrontation with the Netherlands and with Malaysia. In 1965 Sukarno was overthrown, soon to be replaced by President Suharto's administration. In contrast with its predecessor this is considered to have 'pragmatic' economic policies, placing much greater reliance on investment from overseas, on a more 'open' economy for local entrepreneurs and on a process of national economic planning carried out by technically trained economists (Schmitt 1969; Glassburner 1971; Mackie 1971; Hoadley 1978). The national planning agency (BAPPENAS) drew up the First Five Year Development Plan (REPELITA I) which came into operation in 1969 (Japenda 1969; Panglaykim and Thomas 1971), followed in 1974 by REPELITA II (Departmen Penerangan 1974). These are essentially programs of public investment, reflecting a commitment to a stable, mixed economy, avoiding major structural changes, and based on an expectation of a modest rate of economic growth over a long period. However, examination of the real priorities of these plans reveals that most new investments in Indonesia are directed to the prestigious, capital-intensive sector, rather than to the labour-intensive or informal sectors. Equity is being traded off for economic growth (Syamaprasad Gupta 1977: 12-13).

For nearly seventy-five years there has been explicit recognition that over-population in Java and Bali is one of Indonesia's most pressing problems. Many of the inhabitants of these two regions are rural dwellers, and the pressure of population in rural areas encourages migration. While some of this pressure is intended to be relieved by settling people in agricultural development areas in the

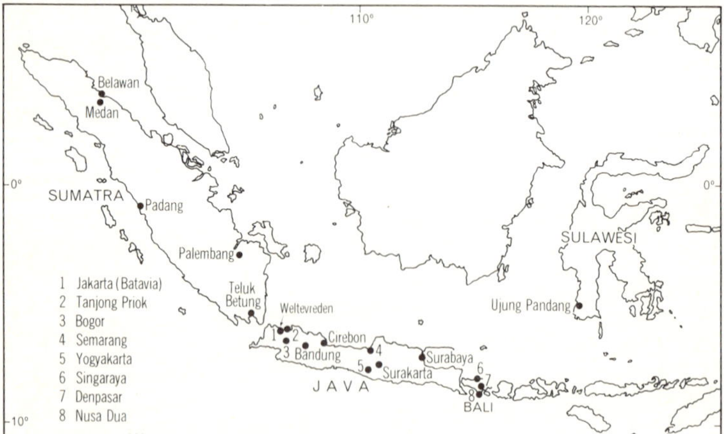

Figure 13.1 Western part of Indonesia. Places mentioned in Chapter 13.

outer islands (Hardjono 1977), many find their way to the cities. In 1971 it was estimated that 5 per cent of the population of Indonesia had migrated from the province of birth (Sundrum 1976). Jakarta, Indonesia's largest city, received 40 per cent of the migrants, and during the decade to 1971 absorbed 400,000 into its work-force (Beier 1976: 390). Technically, since 1970 Jakarta has been a 'closed' city, to which migration is permitted only to those with a job (Sethuraman 1976: 4), but there is little evidence to suggest that the flow of migrants has decreased. By the early 1970s Indonesia had several major urban centres: Jakarta (*c.* 4.5 million); Surabaya and Bandung, Java (over 1.0 million); Medan, Sumatra (over 600,000); Semarang, Java (over 600,000); Palembang, Sumatra (over 500,000).

The Development of Urban Centres in Indonesia

Although large cities are a recent phenomenon in Indonesia, there is a long urban tradition: important centres have existed at least for 1,500 years. Early cities were either foci of political and religious power or mercantile bases for some of the more significant regional trading empires. The political centres had a characteristic morphology with recognisable elements. The ruler resided in a palace within the *kraton* (generally the only part of the city protected by a wall). This area also contained the residences of other important government and religious officials whereas those of minor officials were located just outside. The remainder of the population lived in a series of villages or *kampung* clustered around the *kraton* interspersed with agricultural land (Stutterheim 1948; Cobban 1970: 19-42). Over a long period an intricate and complex set of spatial and social relations evolved between the central locus of power in the *kraton* and the proliferating peri-urban *kampung* communities (van Mook 1958; Moertono 1963). Several such centres have survived although much transformed by their colonial experience; Yogyakarta and Surakarta are examples of rather late manifestations of this style of city (Reed 1976).

Colonial rule by the Netherlands created an additional series of towns. Batavia was founded early in the seventeenth century and, as Dutch power and influence spread, many existing centres were incorporated in a new administrative and commercial network. These cities strongly reflected Dutch interests. Instead of being dominated by traditional rulers and their courts, they functioned as points of impact for a new colonial economy. Dutch commercial and political interests manifested themselves in the form of warehouses, trading facilities, factories, transport facilities, port developments, government and military installations, and residential areas which in their spaciousness of design reflected the grandiose style of living enjoyed by at least a segment of the colonial élite. Batavia's initial site by the coast was considered too cramped and unhealthy; thus early in the nineteenth century the European population gravitated a short distance south to Weltevreden in a response to the move initiated by Governor-General Daendels (Cobban 1971: 149). Later in the century the centre of colonial political power shifted to Bogor (Buitenzorg) but of equal significance was the spread of the Dutch urban network. As the Dutch exercised progressively tighter control over political conditions in the various States and principalities of the East Indies and as Dutch business interests penetrated more deeply into the fabric of Indonesian society both in Java and the outer islands, more colonial type cities appeared. Medan in North Sumatra grew to become a centre for commercial interests related to the burgeoning plantation industry; Semarang, Cirebon and Surabaya became major commercial foci; and Bandung became both a commercial and administrative centre in West Java.

The morphology of these Dutch colonial cities reflected the structure of colonial society although in some instances the *kraton* persisted. The new economic and social modes were quickly reflected in extensive commercial facilities; in Chinatowns where retailing activities predominated (Jackson 1975); in Dutch commercial zones where banks, trading companies, shipping companies, agency houses, and other similar institutions shared space with government offices; in spacious residential areas for the colonial population; and in the dismal, crowded *kampung* where most of the Indonesians lived. Although no evident and conscious town planning program led to this kind of morphology, the political and economic forces operating in Indonesia at the time were themselves sufficient. Urban places were the most important locus of the Dutch presence; Dutch priorities and relationships with other social groups were stamped on the urban landscape.

The early twentieth century saw a crisis in the political and administrative structure of urban Indonesia. Until that time authority within the colonial political structure had been highly centralised, but with the emergence of cities the need for some degree of devolution of responsibilities became clear. A policy of decentralisation adopted in 1903 allowed for the creation of municipalities with powers to raise taxes and provide essential urban services without reference to central government. In practice management of the urban centres proved very difficult. The constitutional arrangements of 1854 had separated the administration of the Indonesian colonies on the basis of ethnic status. The native and Dutch administrations were, in reality, two separate systems: when these interacted in the cities conflict was inevitable and irresolvable. A long struggle ensued during which the cities sought to bring about a unification of urban administrative responsibilities, but the Dutch did not take such a step prior to World War II (Milone 1976: 108-13; Cobban 1970: 117).

Decentralisation helped to encourage the development of town planning. Problems of urban spatial management led to ordinances controlling nuisances (such as noxious industry) and to building codes. But a comprehensive ordinance on town planning was more elusive. Although considered during the 1920s and 1930s, such an ordinance was finally put together in 1938, after town planning had become established practice in European nations. But even this belated legislation, the *Stadsvormingsordonnantie Stadsgemeenten Java,* had to await 1948 before being passed into law.

Developments in the field of town planning since Indonesia gained independence in the late 1940s have been significant although not revolutionary. The original Dutch legislation of 1948 was allowed to stand, but very little action took place in relation to town planning. The Public Works Department undertook some elementary planning responsibilities in the 1950s through a division referred to as the *Balai Tata Ruang* which in the 1960s became the *Direktorat Tata Kota dan Daerah* with responsibility for physical planning at the city and regional levels (Direktorat Jendral Cipta Karya 1973). The 1948 legislation was revised in a series of measures relating to building codes, land use, zoning regulations, and the power of land resumption (McTaggart and Stormont 1975); in addition, the First and Second Five Year Plans have required cities and towns to prepare plans governing their development until at least the end of the century.

Town Planning and the Process of Urbanisation

Town planning has emerged in Indonesia as a form of urban

management. Master plans are generally sectoral plans first and foremost; spatial design is a secondary consideration. Typically a plan document will contain a 'data compilation' amassing all manner of economic and social information about the planning area, which is used to show where deficiencies actually exist, or are likely to exist, during the period for which planning is contemplated. The plan itself is a statement about the required sectoral investment programs if these deficiencies are to be rectified. In some sectors, such as housing, commerce and industry, the investment is expected to come from the private sector. In others, such as health, education, public works and utilities, the public sector will be responsible although the plan itself does not actually make provision for the investment. It is expected that public sector investments through development project channels will be designed to conform to the outlines of the city plan.

Master plans contain provisions relating to the spatial structure of the cities concerned. Zoning schemes are required but, in general, zoning begins by sanctifying present land use patterns. Occasionally new land use elements are provided for — industrial estates for new factory development, or residential areas for 'sites and services development' — and allowance is made for land for major public sector schemes. But existing zoning regulations barely recognise the mixed nature of the functional structure of the Indonesian urban landscape. In spatial terms, master plans incorporate a hierarchical and administrative concept of urban space (McTaggart and Stormont 1975). Most of the major cities in Indonesia have the status of *kotamadya* or municipality, equivalent to the rural *kabupaten* or second level district. The city is thereafter broken down into *kecamatan* and each *kecamatan* into several *lingkungan.* Full-time administrative public servants are employed to staff local offices at these levels. *Lingkungan* are further subdivided into smaller units (commonly *desa,* but the terminology varies). In the eyes of the planners the administrative and functional structures of the city should be brought into conformity with each other. The city, as *kotamadya,* should be a regional commercial, social and cultural centre for the region as a whole, and the city centre should reflect this role. Within the city each *kecamatan,* as a relatively large organisational unit, should contain specialised commercial, health, education and government services. Lower level administrative units should be organised so as to contain less specialised functions —

rudimentary health services, elementary schools and day-to-day commercial establishments. But although this hierarchical concept of urban spatial organisation is given clear expression, it is imperfectly realised in the implementation of plans themselves.

Plans make little or no reference to some of the most fundamental characteristics of the Indonesian city. The city is the locus of contact between the 'monetised' international segment of Indonesia's economy, and its more 'traditional' segment. Just as cities were the primary loci of economic activity and management during the Dutch period, so today they reflect the import and export of commodities, the channelling of trade between capitalistically organised entities of concentrated economic power and the diffuseness of a peasant style economy. The dichotomy is not one of 'urban' versus 'rural' but rather of two systems in which both the mode of production and production relations differ (Franklin 1965). The 'monetised' sector is represented not only by industrial, commercial and financial activities located in the cities, but also by plantation, mining and construction activities in rural areas. The 'traditional' sector is represented by subsistence and semi-subsistence activities in agriculture as well as the vast 'informal' sector in the cities; the multitudes who scrape a precarious living by hawking, pedalling *becaks,* scouring garbage dumps, picking up cigarette ends and other similar activities (Santos 1971; McGee 1974). Planners recognise the fact of poverty, but not the dichotomy of production relations.

Inevitably, then, cities are planned and managed primarily in terms of the economic logic of the 'monetised' sector, in which powers of control and articulation are much more strongly expressed. The city has become not merely a locus of economic activities important for the 'monetised' sector but an integral part of these economic activities. The city — the land and its buildings — becomes a category of commodities; the city becomes a place where people are conditioned to live according to the economic principles of a capitalistically organised society, and also becomes an entity where public sector investments are channelled as much for the benefit of 'monetised' sector interests as for the advantage of the populace at large. Here it is intended to concentrate attention on four aspects: the manner in which the organisation and development of a city provide external economies which are of substantial benefit to the 'monetised' sector; the effect of urbanisation in creating a concentrated labour force; the role of urban life in establishing a market, however small the individual's purchasing power may be, for the products of the 'monetised' sector; and the impact of urban growth of property ownership.

Planning and external economies

There are many examples of projects proposed for Indonesian cities in the First and Second Five Year plans which were calculated to improve the external economies of firms. There are several important developments in transportation. The port of Tanjong Priok (Jakarta), after years of comparative neglect, is being doubled in capacity (Government of Jakarta 1967: 41-4); the port of Belawan is to be developed and linked to Medan by an 'industrialised corridor' (Kotamadya Medan 1973); Bali is being provided with a new port and township (BAPPEDA Bali 1977); regional planning in South Sumatra hinges upon the growth of Teluk Betung, which is intended to function as a development pole (Bonn Universität 1974); and Padang has been selected to become the major port for the whole of the western part of Sumatra (Kotamadya Padang 1974).

Improvements in road transport facilities will also provide external economies to the monetised sector firms. During the colonial period investment in street development was mainly in the Dutch business, administrative and residential areas. The *kampung* areas were served only by narrow unsurfaced paths which were quickly churned to mud during rain. There are numerous small-scale schemes in Indonesian towns to improve the quality (but not width) of *kampung* access paths. In contrast major developments are being carried out on arterial links which serve the commercial and the better residential areas. In Jakarta, Jalan Thamrin was built initially in the Sukarno period as a business thoroughfare but subsequently has had to be almost completely rebuilt. Jakarta and Bogor are being connected by a fast arterial which will undoubtedly become a major axis of Indonesia's most developed industrial zone. Medan proposes to build a series of ring roads to improve traffic flow around the city (Kotamadya Medan 1973).

External economies are also provided in the form of industrial sites. During the colonial period industrial location was constrained only by the *Hinderordonnantie* legislation dealing with noxious industry. Current master plans are generally more positive with industry being recognised as a zoning category. In several instances there are provisions for the development of planned industrial estates (Scott 1971).

Labour flow

Despite Indonesia's large population, and despite the preoccupation with population growth rates, provision of labour to monetised sector activities remains a problem. Large-scale enterprises in the colonial period very often had to effect their own migration schemes: plantation workers in Sumatra were recruited from Java, and labourers for coal, tin and petroleum operations from even further afield. Contemporary industrial development demands skilled and disciplined labour, qualities that industrialists believe cannot normally be found among the swelling masses of the urban 'proto-proletariat' (Papanek 1975). Public sector investments have therefore been directed towards vocational and industrial training. Thus Bandung has become the nation's largest textile manufacturing centre as well as a major centre for teaching textile technology. Similarly, Surabaya trains marine engineers. In Bali, where tourism is now the major industry, government provided the initial tourism training centre in Denpasar, a function which is to be transferred to Nusa Dua as part of a large-scale tourist development project being undertaken with World Bank support (McTaggart 1978a).

The urban proto-proletariat functions, nonetheless, as a vast pool of low-cost, unskilled labour. Most of the migrants to the city led lives of great poverty in the villages, but find only poverty in the cities (Montgomery 1975; Penny and Singarimbun 1973; Temple 1975). They depend on urban employment in the 'informal' sector — most of which is discontinuous and unremunerative. Sethuraman (1976: 13) estimates that over 500,000 people were employed in the 'informal' sector in Jakarta in 1971, four times as many as worked in manufacturing. Industrial and commercial developments in the city can therefore count on easy availability of labour, even if it is of relatively low quality. Workers can be employed as and when needed and then thrown back upon the 'informal' sector, which thus acts in a social security role at little or no cost to the 'monetised' sector firms.

Not surprisingly planners adopt an ambivalent attitude towards the 'informal' sector. For example, many such activities gravitate around the municipal markets of Indonesian towns. Although much illegal selling takes place in surrounding streets, the markets themselves provide opportunities for low-cost stalls to operate and thus effectively participate in the 'informal' sector. *Becak* transport has roots in both the 'monetised' and the 'informal' sectors: equipment for hire is provided by capitalists but the riders themselves are best considered to be part of the 'informal' sector. Most cities tolerate some freedom of operation for *becak* although they may have to carry a licence and be limited as to the streets and distances they can cover. Yet their numbers are seldom controlled, and there is little interference in the commercial arrangements riders make with owners (Forbes 1978). Planning documents scarcely ever refer to informal activities such as hawking or the operation of *warung* (sales kiosks which are often set up in homes or compounds); sometimes such endeavours are suppressed when they cause congestion or become a nuisance. Labour management and organisation *per se* are not generally a part of urban planning but the plans operate in a context in which one of the functions of the city is to sustain a labour pool at a level of living approaching subsistence.

Urban areas as a consumer market

Indonesian exports are predominantly from rural areas. Petroleum, tin, rubber, palm oil, tropical timber, coffee, nickel and many others are mainly derived from outer islands away from the urban centres. Industry at present is concerned with import substitution rather than with export markets. Thus several aspects of Indonesian commercial development are directed towards enhancing consumer purchasing power, however limited this may be, especially in the urban areas, where retail distribution is better articulated.

A central objective of the Indonesian government has been to control the price of basic foods, especially rice, in the cities. Many government and military employees were, for a long time, paid partly in rice, and such forms of remuneration continue. The Indonesian government also maintains a centralised rice acquisition program and closely controls imports of rice. Substantial improvements in rice production in Indonesia have been achieved during the past decade, but the benefits have apparently not accrued to the rural peasantry but rather to land-owners and urban consumers (Franke 1972; Krinks 1978).

Although urban plans reveal no uniform housing development policies, efforts are made to avoid increases in housing costs for the bulk of the population. Property taxation rates are maintained at an extremely low level (McTaggart 1976). Public housing is not being widely constructed anywhere in Indonesia, but 'sites and services' schemes are being encouraged (Direktorat Tata Kota dan Daerah 1974). Urban transport costs are also kept low, mainly by permitting unrestricted competition. As basic living costs remain modest, most urban households can afford some purchases beyond their most elementary needs; collectively the impact of this large, though poor, consumer market is of considerable significance.

Planning and urban property ownership

Conflicting motives and policies are also evident in urban land policies. The interests of the 'monetised' sector are best served when the cities act as powerful agents of capital accumulation through a rise in the values of marketable property; yet such a trend undermines the policies of maintaining a large low-wage labour force and reasonable consumer purchasing power. Urban planning has generally facilitated the expansion of a market in urban property but, so far, a rapid and uncontrollable rise in property values has not been a feature of urban Indonesia except in and around Jakarta.

Land tenure in Indonesia is extremely complex. Traditional tenure systems vary between regions, and the complexities were not diminished by Sukarno's attempt to unify and codify Indonesian land legislation in 1960. The *Undang Undang Pokok Agraria* of that year provided categories of land title which conformed to Western notions of ownership and leasehold, but there was no compulsion for people to convert their property from traditional tenures to those provided for in the legislation, and in practice, such conversions have proved to be difficult and tedious. City core areas, as they existed in the Dutch period, have long been under Western forms of tenure. However, the fringes and the zones of *kampung* housing have remained under traditional forms: as these areas become integrated into cities the problems associated with the lack of marketable title increase.

Urban development and planning have proceeded on the assumption that marketable titles would prevail. The expected dynamics of urban land use and change entail exchange of property through a market system which will reflect 'normal' location and 'highest bidder' rent factors, but the land tenure situation fetters its operation. In Padang, much of the land now coming into the urban area remains under Minangkabau group tenure (Evers 1974a,b); decisions made about its use have to be referred to all interested parties — an almost impossible task. In Bali substantial areas are held not by householders but by the *banjar* (the village community). As individuals cannot make decisions about the use of such land planners are unable to exclude non-urban activities from their schemes; as urban places expand they encompass areas which continue to be used for farming because there are no ready means to compel or encourage their conversion to urban uses (McTaggart 1978b).

Questions of land tenure also underlie attitudes towards squatter settlements, a common feature in Southeast Asian towns. Illegal occupation of land by impoverished migrants and other urban dwellers has long been a feature of Indonesian cities. In Dutch times there was only partial control over the development of *kampung* areas, and the European-dominated city governments never achieved real authority over the native *kampung* within their own city limits. During the early years of independence political and military insecurity in rural areas contributed to a large flow of migrants to the towns, and many of these had no choice but to occupy whatever space they could find. Around Medan in North Sumatra whole plantations were invaded and taken over by squatters. In Ujung Pandang, Sulawesi, large tracts of unalienated land around the city were occupied and, although much of this land is now technically owned by the state, its occupants have been recognised as *de facto* owners and may even have some kind of documentation to sustain their claim (McTaggart 1976).

Indonesian planners have generally shared the normal hostility towards the squatters, who are seen as an obstacle to orderly and healthy development. In practice this hostility is muted. In Jakarta frequent sweeps are directed by the police against the squatters whose feeble dwellings mushroom overnight along railway tracks and river banks. But such hardline policies are not characteristic of other cities, and if squatter communities police and contain themselves they are left in peace. Squatters form no part of city planning schemes in Indonesia: various forms of lower cost housing are sometimes included in zoning provisions, but specific measures to deal -with illegal occupance are absent.

Planning thus not only sidesteps major issues relating to land tenure in Indonesian cities but some of its basic assumptions imply the functioning of a property market along the lines of a Western city. It has been indicated already that master plans make no provision for the investments needed for their realisation. Public sector investments stem from decisions made by various levels of governmental bodies, and planning is intended to help guide decision-making. Private sector investment is expected to adhere to the outlines of the plan which, it is believed, provides for an attractive private investment climate in urban real estate that will stimulate adequate housing and commercial construction. So far, however, this assumption remains unproved. As suggested previously, speculative conditions have developed in Jakarta where effective demand for housing from a growing middle class and expatriate community creates a distinctive situation, but other major cities, too, are now beginning to feel the pressures of speculative demand. Institutional structures, however, are inadequate to sustain a general rise in the tempo of residential construction in Indonesia. Government policy is not to allow much capital to find its way into 'social' as opposed to 'productive' investments, and a massive upgrading of housing through high cost construction is probably still many years away. Planning seeks to create market conditions for urban property but the necessary financial infrastructure to sustain such a market is not yet in sight.

Planning and the Ethos of 'Urbanism'

No review of planning would be complete unless it raised the question of how planners approach one of Indonesia's central social problems — inter-ethnic relations. Since the advent of the Netherlands as a colonial power and the active development of colonial cities, mixing of ethnic groups has been one of the most pronounced features. Just as cities in medieval Europe served as places of refuge where feudal bonds no longer applied, cities have served as places where Indonesians could elect to live outside the political system of their native location. Even though the Dutch finally decided in 1854 to create a separate native administration embracing those Indonesian natives living in the urban areas, this never operated as a fully traditional political system. Freed from traditional constraints Indonesian natives were able to interact with immigrant Dutch, Chinese, Arabs and other diverse ethnic elements, including those from other parts of the archipelago with which contacts were seldom made (van der Kroef 1952; Castles 1967).

The city quickly became a place both of contact and of separation of different peoples. It provided an arena for the interaction of the various ethnic groups, the so-called 'plural society', mediated not only by the economic and market factors so obvious to Furnivall, but by the whole colonial social structure. A degree of distinctness was maintained, in which transformed ethnic elements confronted each other, keeping their identities by means of a greatly modified form of an ancestral culture. Colonial Dutch, *mardijkers,* Indonesian migrants and Chinese all clung to and were compelled to retain a sense of separateness which was quickly translated into distinctive styles of life and even landscape elements. The Dutch authorities assigned themselves spacious residential areas through a combination of political and economic influence (Cobban 1971); native Indonesians sought out an appropriate *kampung* (Cobban 1970); and Chinese were long restricted to defined quarters which still contain a large proportion of the Chinese commercial population of the cities (Jackson 1975; McTaggart 1976, 1978b). The Dutch have gone; the Chinese are no longer defined as a legal entity though some of them have declined to adopt Indonesian citizenship. But despite the national unification policies followed by successive Indonesian governments since the end of World War II (Lev 1973), ethnic differentiation is still a major preoccupation.

Master plans generally decline to recognise overtly any form of ethnic separation yet dynamic urban processes frequently revolve around precisely such issues. During the period of rapid in-migration into Indonesian cities in the 1950s and 1960s ethnic communities almost automatically appeared (for example see Bruner 1972 for *Batak* people in Medan, Sumatra). In Singaraya, Bali, it has been observed that religion becomes a fundamental principle in the organisation of communities; the incoming Javanese Moslems and the native Balinese Hindus, each strive to associate with a *desa* of their own religious affiliation. Urban plans generally recognise group and community separation by class differentiation with residential neighbourhoods sometimes being distinguished as 'high-cost' or 'low-cost'. Only in exceptional cases do plans allow for the cultivation of distinctive ethnic character in component parts of a city.

Conclusions

Although town planning is usually considered to be a conscious activity, reflecting the overt desires of an administrative or planning system, frequently its most important effects are involuntary and its most fundamental assumptions unspoken. Traditional Indonesian urban forms were a reflection of the cosmology of the times, regardless of the views of the craftsmen and artisans who built them. In the colonial period the Dutch planned their cities for safety and comfort, as well as for an appropriate aesthetic quality; their efforts reflected the ideology of colonialism. The contemporary city is planned as a centre for modernisation; it can be viewed as an expression of capitalism in the organisation of living space.

The economic and social philosophy of the New Order government in contemporary Indonesia is to galvanise society into progress and development. New habits and attitudes have to be encouraged as well as new forms of labour and production. Since this new 'pragmatism' is based on Western models, new standards of living space are also required. A modern city is a symbol of a modern society, and from it will diffuse modern ideas and modes of organisation. Planning should perform this role and help provide the nation with cities of which it can be proud. At this stage such a picture of Indonesian development remains over-optimistic. Development policies have encountered social obstacles as well as those related to lack of capital or natural resources, because of a resistance to the transformations which a 'modern' economy evidently requires. Similarly there is, or would be, resistance to the implementation of a form of city planning which is very much the handmaid of the new economy. Planning, however, remains only partly effective, with many of its provisions never to be realised, and many others to be greatly modified before being put into effect. Nonetheless, the very institution of planning is both a reflection of the establishment of capitalistic modes of thinking and an encouragement to their further development. Indonesia is planning for a specific economic system rather than for an abstract quality in the development of its urban centres.

References

Allen, R. 1975. The city is dead. *The Ecologist* 5: 186-9.

BAPPEDA Bali 1977. *Pra-konsep Master Plan Wilayah Pelabuhan Celukan Bawang*. Denpasar.

Beier, G. *et al.* 1976. The task ahead for the cities of developing countries. *World Development* 4: 363-409.
Bonn Rheinische Friedrich-Wilhelms Universität 1974. *Sumatra Regional Planning* Study. Direktorat Tata Kota dan Daerah, Jakarta.
Breese, G. 1966. *Urbanization in Newly Developing Countries.* Prentice Hall, Englewood Cliffs.
Bruner, E.M. 1972. Batak ethnic associations in three Indonesian cities. *Southwestern Journal of Anthropology* 28: 207-29.
Castles, L. 1967. The ethnic profile of Jakarta. *Indonesia* 3: 153-204.
Cobban, J.L. 1970. *The City on Java: An Essay in Historical Geography.* Ann Arbor, University Microfilms.
Cobban, J.L. 1971. Geographic notes on the first two centuries of Djakarta. *Journal of the Malaysian Branch of the Royal Asiatic Society* 44(2): 108-50.
Departmen Penerangan 1974. *Rencangan Pembangunan lima tahun 1974/5 - 1978/9.* Jakarta.
Direktorat Jendral Cipta Karya 1973. *Programs and Approaches to Promote Physical Planning for Regional Development in Indonesia for the Second Five Year Plan.* Jakarta.
Direktorat Tata Kota dan Daerah 1974. *Sites and Services.* Jakarta.
Evers, H.-D. 1974a. Traditional land tenure in an Indonesian city. *Land Tenure Center Newsletter.* No. 43: 14-19. University of Wisconsin, Land Tenure Center.
Evers, H.-D. 1974b. *Changing Patterns of Minangkabau Urban Landownership.* University of Singapore, Department of Sociology, Working Paper No. 22.
Forbes, D. 1978. Urban-rural interdependence: the trishaw riders of Ujung Pandang. Pp. 219-36 in P.J. Rimmer, D.W. Drakakis-Smith and T.G. McGee (eds.), *Food, Shelter and Transport in Southeast Asia and the Pacific.* The Australian National University, Department of Human Geography Publication HG/12(1978).
Franke, R. 1972. Theories of the green revolution. (Paper read at American Anthropological Association Meeting, Toronto.)
Franklin, S.H. 1965. Systems of production; systems of appropriation. *Pacific Viewpoint* 6: 145-66.
Glassburner, B. 1971. Economic policy making in Indonesia, 1950-1957. Pp. 70-98 in B. Glassburner (ed.), *The Economy of Indonesia: Selected Readings.* Cornell University Press, Ithaca.
Gomhan, V.M. *et al.* 1976. The economic geographical aspects of world urbanization and its specific features in the developing countries. Pp. 25-45 in A. Manzoor and B. Pokshishevsky (eds.), *Urbanization in Developing Countries.* Osmania University, Hyderabad.
Government of Jakarta 1967. *The Master Plan of Jakarta.* Jakarta.
Hardjono, J.M. 1977. *Transmigration in Indonesia.* Oxford University Press, Kuala Lumpur.
Harvey, D. 1973. *Social Justice and the City.* Arnold, London.
Hoadley, S.J. 1978. The politics of development planning agencies: the evolution of Indonesia's Bappenas. *Asia Quarterly* 39-66.
Jackson, J.C. 1975. The Chinatowns of Southeast Asia: traditional components of the city's central area. *Pacific Viewpoint* 16: 45-77.
Japenda 1969. *Development: Summary of Indonesia's First Five Year Plan, April 1 1969-March 31 1974.* Jakarta.
Kotamadya Medan 1973. *Garis besar kebijaksanaan penyusunan master plan, Kotamadya Medan.* Medan.
Kotamadya Padang 1974. *Rencana pembangunan 1974/5-1978/9.* Padang.

Krinks, P.A. 1978. Rural changes in Java: an end to involution? *Geography* 63: 31-6.
Lev, D.S. 1973. Judicial unification in post-colonial Indonesia. *Indonesia* 16: 1-38.
McGee, T.G. 1974. *The Persistence of the Protoproletariat: Occupational Structures and Planning for the Future of Third World Cities*. U.C.L.A., School of Architecture and Urban Planning, Los Angeles.
McGee, T.G. 1976. Beach-heads and enclaves: the urban debate and the urbanization process in Southeast Asia since 1945. Pp. 60-75 in Y.M. Yeung and C.P. Lo (eds.), *Changing South-East Asian Cities: Readings on Urbanization*. Oxford University Press, Singapore.
Mackie, J.A.C. 1971. The Indonesian economy, 1950-1963. Pp. 16-69 in B. Glassburner (ed.), *The Economy of Indonesia: Selected Readings*. Cornell University Press, Ithaca.
McTaggart, W.D. 1976. Urban policies in an Indonesian city: the case of Ujung Pandang, South Sulawesi. *Town Planning Review* 47: 56-81.
McTaggart, W.D. 1978a. Tourism and tradition in Bali. (Paper read at Western Conference of the Association for Asian Studies, Tucson, Arizona.)
McTaggart, W.D. 1978b. Land use in Sukabumi, West Java. (Paper read at Meeting of Association of American Geographers, New Orleans.)
McTaggart, W.D. and Stormont, D. 1975. Urbanization concepts in the restructuring of Indonesia. *Journal of Tropical Geography* 41: 34-44.
Milone, P.D. 1976. Contemporary urbanization in Indonesia. Pp. 91-9 in Y.M. Yeung and C.P. Lo (eds.), *Changing South-East Asian Cities: Readings on Urbanization*. Oxford University Press, Singapore.
Moertono, S. 1963. *State and Statecraft in Old Java: A Study of The Later Mataram Period, 16th to 19th Century*. Cornell University, Department of Asian Studies, Ithaca.
Montgomery, R.D. 1975. Migration, employment and unemployment in Java: changes from 1961 to 1971 with particular reference to the green revolution. *Asian Survey* 15: 221-33.
Panglaykim, J. and Thomas, K.D. 1971. *Economic Planning Experience in Indonesia*. Nanyang University, Singapore, Institute of Business Studies, Occasional Paper No. 5.
Papanek, G.F. 1975. The poor of Jakarta. *Economic Development and Cultural Change* 24: 1-27.
Penny, D.H. and Singarimbun, M. 1973. *Population and Poverty in Rural Java: Some Economic Arithmetic from Sriharjo*. Cornell University, International Agricultural Development Monograph No. 41, Ithaca.
Redfield, R. 1953. *The Primitive World and its Transformations*. Cornell University Press, Ithaca.
Redfield, R. 1956. *Peasant Society and Culture: An Anthropological Approach to Civilization*. University of Chicago Press, Chicago.
Reed, R.R. 1976. Indigenous urbanism in South East Asia. Pp. 14-27 in Y.M. Yeung and C.P. Lo (eds.), *Changing South-East Asian Cities: Readings on Urbanization*. Oxford University Press, Singapore.
Santos, M. 1971. *Villes du Tiers Monde*. Techniques (Géographie Économique), Paris.
Schmitt, H.O. 1969. *Monetary Policy and Social Conflict in Indonesia*. University of California Press, Berkeley.
Scott, W.D. and Co. Pty Ltd 1971. *Report on Preliminary Survey of the Development Potential of the City of Surabaya in Particular and East Java Province in General with Special Reference to Industrial Estates*. Surabaya Investment Information Office.

Sethuraman, S.V. 1976. *Jakarta: Urban Development and Employment.* International Labour Office, Geneva.

Stutterheim, W.F. 1948. *De Kraton Van Majapahit.* Martinus Nijhoff, 'S-Gravenhage.

Sundrum, R.N. 1976. Inter-provincial migration. *Bulletin of Indonesian Economic Studies* 12(1): 70-92.

Syamaprasad Gupta 1977. *A Model for Income Distribution, Employment, and Growth: a Case Study of Indonesia.* Johns Hopkins University Press, Baltimore.

Temple, G. 1975. Migration to Jakarta. *Bulletin of Indonesian Economic Studies* 11(1): 76-81.

van der Kroef, J.M. 1952. The Indonesian city: its culture and evolution. *Asia* 2: 563-79.

van Mook, J. 1958. Kuta Gede. Pp. 275-331 in W.F. Wertheim (ed.), *The Indonesian Town: Studies in Urban Sociology.* van Hoeve, The Hague and Bandung.

Weber, A.F. 1899. *The Growth of Cities in the Nineteenth Century: A Study in Statistics.* Macmillan, New York.

Weber, M. 1958. *The City.* The Free Press, Glencoe, Ill.

Wirth, L. 1938. Urbanism as a way of life. *American Journal of Sociology* 44: 1-24.

employed on the holdings, large or small, of other nationals. These data would be virtually impossible to obtain under present conditions; their lack is among the factors contributing to 'concealment' of the extent of such employment. The total number of agricultural wage earners probably does not exceed 60,000, an estimate which allows for many who are seasonal or temporary workers. Nearly 70 per cent of this estimated total are employed on 'largeholdings' (Table 10.1). Although a seemingly small number, 60,000 represents over 30 per cent of the monetary sector work-force, and about 90 per cent of the unskilled workers in rural wage employment.

The decline in the number of largeholdings, and in the work-force employed on them, during the 1970s is apparent from Table 10.1. The 27 per cent decline in the number of largeholdings was due to a combination of factors which included amalgamation of some properties, resumption of some plantations for a new smallholder oil palm project in Northern Province, and the transfer of a number of holdings from expatriate to national ownership under the Plantation Redistribution Scheme (see later). The work-force on the largeholdings declined by about 20 per cent during the same period. Of the 10,000 employees who left, it is likely that some returned to their villages, some migrated to urban centres, and others found employment with village big-men elsewhere in the rural sector. Rising labour costs during this period led to the substitution of plantation workers by village labour contracted for specific tasks such as harvesting on a number of largeholdings (R. F. McKillop, personal communication 1979); such temporary labour is unlikely to appear on employment returns submitted by plantation managers.

In general, the rural wage workers come from the Southern Highlands, Enga and Chimbu Provinces, more remote regions of the Eastern and Western Highlands, and parts of the Sepik Provinces. Inland peoples in coastal provinces, such as Madang, Morobe and Northern, also form part of this work-force (Fig. 10.1). The employment areas, broadly, are the plantation districts and regions where the availability and suitability of land allow extensive peasant cash cropping. In many areas agricultural workers are employed on the holdings of big-men within the same province, but cases are also known of long distance migration for such employment. Rowley (1958: 542) commented on the 'village dwellers who are full time wage earners' and 'workers from distant areas, some living with their

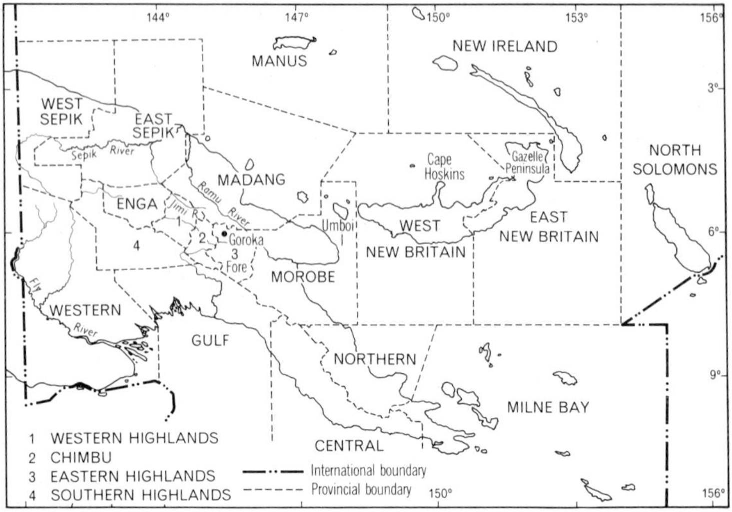

Figure 10.1 Papua New Guinea. Provinces and places mentioned in Chapter 10.

families on the job, who have forgotten their village ties'. In 1960, only a decade after the establishment of expatriate coffee plantations in the Eastern Highlands, some villagers near Goroka were employing men from the Fore region (Howlett 1962: 193-9). By 1978 this was a widespread practice in villages around Goroka, and the agricultural workers, some with their families, came from various parts of the province and also from Chimbu. According to Heaney (1977: 5),

> ex-plantation workers, from Chimbu, the Southern Highlands and the Jimi Valley were laborers for Wahgi big-men and preferred to work for less pay than they would receive on nearby plantations. Informants stated that working as a client had certain social advantages to being a wage-earner on a plantation.

All four New Ireland villages surveyed by Young (1977) included men from other provinces who had permanently resettled (in the case of some Sepik men, before World War II). Some were employed by villagers with large areas of land. Connell (1977: 29) reported that on Bougainville, 'some Siwais have imported New Guinean labourers'. Grossman and Ploeg, in separate personal communications, have observed that hired labour is used by highland cattle owners, by Tolai big-men and by Umboi Island people. Ploeg thought that a similar situation would arise in the resettlement scheme at Cape Hoskins, where cases tantamount to absentee landlordism have appeared.

The widespread evidence of increasing agricultural wage employment in villages is to some extent a reflection of the significant changes which have been made both in the pattern of plantation ownership and in rural employment legislation during the 1970s. With the attainment of self-government and independence, the adoption of the Eight Point Plan, the announcement of citizenship criteria, and the establishment of the government's Plantation Redistribution Scheme, a number of foreign-owned plantations have been acquired by local groups. Probably very few individual Papua New Guineans owned plantations established on formally-alienated land during the colonial period. Crocombe (1965: 35) noted that until the early 1960s 'the possibility of indigenous take-over of existing expatriate plantations was not considered'. He mentioned the 'extreme reluctance' of the former Native Loans Board to grant loans for the employment of labour, and that exceptions to the granting of family-sized blocks were rare. Australian policy was to lease largeholdings to expatriates and smallholdings to nationals. However, some expatriate plantations were acquired by co-operative societies or communal groups during the 1960s (Crocombe 1965; McGregor 1971), and some extensive plantings of cash crops have been made by nationals on customary land.

An important factor underlying the establishment of the Plantation Redistribution Scheme was that the country 'was embarking on independence with a dominant and pervasive sector of its economy under the ownership and control of foreign interests', but since the late 1960s, in anticipation of independence, 'capital expenditure on plantation management and expansion was in decline and crop volume and quality were deteriorating' (Fingleton 1977: 1-2). Fingleton (1977; in press) has reviewed the background and early operation of the Scheme, which arose from land law reforms proposed by a Commission of Inquiry into Land Matters appointed in 1973. Most plantations acquired under the Scheme have been transferred to the traditional owners of the land, but the Scheme also created a climate which encouraged the purchase of plantations by other groups of nationals, by local government councils and, more recently, by some provincial governments. Some purchasers have formed investment or development corporations for the purpose. The Scheme has provided financial assistance for the purchase of about fifty properties, but perhaps another 100 or so have been transferred from expatriate to national ownership through finance provided by such institutions as the Papua New Guinea Development Bank and the Papua New Guinea Banking Corporation.

Unexpected problems arose during the Scheme's first few years of operation. A first priority was the acquisition of plantations in regions which were economically depressed or in which land was short but, in the absence of clear guidelines for identifying groups eligible for assistance, a number of properties was acquired 'in other areas than was originally intended . . . under pressure from leading national politicians' resulting in 'serious distortion' of the Scheme's intended operation (Fingleton 1977: 8-9). McKillop (in press) has identified another form of distortion of the Scheme's intent:

> Many prosperous groups assuming the control of plantations have maintained a labour force from disadvantaged areas of the country, sought government assistance to provide management advice and training, begun planting more tree crops in areas of acute land shortage, and prevented villagers with rights to the land from 'squatting' on the plantation.

This occurred particularly on the Gazelle Peninsula, where shortage of land was the result of extensive alienation for plantations early in the colonial period. Meanwhile, those groups short of land because of population pressure or conditions inhospitable to cash cropping have been neglected, the gap between the privileged minority and the rest has been widened and regional inequalities have increased. Circumstances such as the boom in coffee prices prior to the steep decline in mid-1977 have accentuated this trend. Thus, one highland group was not only able to repay within a year a government loan it had obtained to purchase a plantation, but was assisted by the local council to build roads and provide a water supply and by the Alienated Land Resources Branch to manage its finances (*Post Courier* 10 December 1976). One locally owned development corporation increased its capital fivefold in twenty months (*Post Courier* 1 February 1977). In 1978 three tribes in the Wahgi Valley were able to raise a loan of 2.3 million kina ($A2.7 million) to buy four coffee plantations. For one tribe the minimum individual shareholding was K400, and 'only a few' were unable to raise this amount (*Sydney Morning Herald* 22 July 1978).

McKillop (in press) contends that the very maintenance of the plantation system in Papua New Guinea perpetuates an imperialist model 'founded on a rigid class distinction between managers and labourers'. Smith (1977: 237) has drawn attention to Gilbert's (1974) work in Latin America, which stressed the role of *internal colonialism* as an aspect of the economic and social structure through which the privileged exploit the poorer classes. A similar situation appears to be developing in Papua New Guinea. On present trends the country's population will have more than doubled by the end of the century, and there is little evidence of increased productivity in subsistence agriculture. In a situation of limited and increasingly scarce resources the growth of a rural élite suggests that other sectors of rural society will inevitably suffer relative disadvantage in access to local resources (especially of land) and to government assistance. Thus in parallel with the growth of a rural élite the extension of marginalised, land-short groups may be expected. One of the few options remaining to such people is wage labour.

Labour problems were another factor which in the early 1970s led to a significant break with the past in the rural sector. Throughout most of the colonial period, the labour market was in a state of perpetual crisis. The situation which Rowley (1958: 546) described in the late 1950s, 'migrant labour — inefficiency — low wages and poor management — rejection of wage earning — more migrant labour', had existed for a long time, and had deteriorated a decade later. Two conferences to assess the Highland Labour Scheme were held in 1966 and 1969. Reports on rural wage labour published in 1970, known as the Cochrane Report (Territory of Papua New Guinea 1970a) and the Isaac Report (Territory of Papua New Guinea 1970b), and the introduction in 1971 of the Native Employment (Minimum Conditions) Ordinance, were also crucial in producing a more receptive attitude to the eventual growth of a permanent rural work-force. The Cochrane Report (1970a: 63) 'addressed itself to the question of how best to promote the further development of a committed wage earning force able to take care of itself'. Densley (1972: 334) noted that:

> The overall concept behind all [its] recommendations is the transition of the wage structure from one of relatively low wages and low productivity per worker and where villagers retain close links with village life to one of higher wages and higher productivity per worker and with a more permanent shift in the work-force. This it envisages as being achieved through the employment of more married workers resident on properties with their families.

As a result of these reports rural wages were increased and a minimum wages tribunal established.

The amended employment legislation of 1971 changed the repatriation provisions for agreement workers, ending the compulsion upon employees to return home if they did not wish to do so. Thus the employment legislation, which had previously been a vehicle for the provision of migrant labour, opened the way for permanent rural wage employment (Smith 1975). The 1966 conference on the Highland Labour Scheme unanimously agreed that the Scheme should be perpetuated as a source of labour for employers, and the Isaac Report considered that terminating the agreement system at short notice would result in a surplus of labour seeking employment, the probable drift of large numbers into urban areas and a considerable reduction in plantation output; however, the Cochrane Report concluded that the agreement scheme, including the Highland Labour Scheme, should be terminated.

In 1974 the Highland Labour Scheme was terminated and replaced by the Rural Employment Programme which 'removed many of the restrictions of the old system and established common provision for the recruitment of agreement labour from the highlands and from coastal areas' (May and Skeldon 1975: 7). A new Employment Act was introduced in the National Parliament in August 1978 but made no major departures from the legislation of twenty years earlier, apart from incorporating those already noted. Most provisions of the legislation, including the minimum wages, are unknown by employees and either unknown or ignored by village employers. The rural minimum wage from 1 September 1978 was K10.98 per week, but many wage workers in villages receive only half this amount. As increasing numbers enter village, as opposed to plantation, employment it seems likely that relations between employers and employees may become more flexible, with a wider range of accommodation between the two. The proportion undertaking agreement labour will probably decline unless workers attempt to protect themselves by insisting on agreements, but in areas where the supply of labour exceeds the demand, minimum wages and other provisions of the law may be forfeited by workers for other considerations. Heaney (1977: 5) found that some workers in villages preferred the shorter and more flexible working hours, and the possibility that their children might become members of the big-man's clan.

Before the Rural Employment Programme replaced the Highland Labour Scheme the government carried out a survey of sixty-six plantations to assess the pre-conditions necessary to encourage a move away from the use of migrant labour in favour of a more settled married work-force in the plantation sector (Joel 1974: 9). Ten of the survey plantations were selected as a control group because they were known to have work-forces consisting mainly of family units more or less permanently settled, although the location of these plantations was not indicated. The survey concluded (Joel 1974: 37) that 'it is relatively cheaper to operate plantations already having settled married work-forces than those still concentrating in employing individual workers such as casual, HLS and verbal agreement workers'. If widely adopted on the remaining expatriate plantations, this finding is likely to give considerable impetus to rural proletarianisation. A settled married work-force was considered to be both less expensive and more productive than the agreement system which has always involved a high turnover of single workers. The study assumed that the displacement from plantation employment of younger single men would 'greatly boost' economic and social development in their home areas, because plantation recruitment has led to a marked slow-down or complete standstill in such development. However, this seems to be confusing cause and effect in many areas. It would appear that young men migrate precisely because their home areas (the recent source areas of agreement labour) have so little economic potential. A consequence different from that envisaged by Joel may follow if plantations adopt a preference for a settled work-force, namely permanent rather than circular migration of men from their home clans.

Although the discussion has been couched in terms of the growth of a rural proletariat, an appropriate typology by which to characterise the quasi-peasant and post-peasant rural people in Papua New Guinea is by no means clear. Mintz (1974: 201) observed that the rural proletariat is 'a typological category that has received relatively little attention in the literature of the social sciences, in spite of its importance — particularly in colonial and neo-colonial societies'. Rural societies in developing nations are varied and dynamic, as attested by the efforts of those who have attempted to construct typologies. Mintz (1973: 92) wrote somewhat ruefully, 'Debates about who peasants are, or how to define peasantries . . . promise to be unending'. Franklin (1962: 1) noted that 'anyone who takes up the study of peasant life soon discovers for himself the paradoxical quality of that existence', and Shanin (1977: 1939) pointed out that 'while different schools multiply argument and terminology, none has managed to avoid the issue of peasant differentiation and structural change'. Perhaps for this reason the peasantry has been termed 'the awkward class' and 'not a class but a notion'. Thus a variety of viewpoints is available. Shanin (1971: 296-8) concluded that a number of 'analytically marginal groups', including agricultural labourers, frontier squatters and pastoral tribal people, must be included with the 'hard core' as elements of increasingly differentiated peasant societies, and affirmed this position in a subsequent paper (1973: 72). Mortimer (1975: 12) took a similar stance in including landless labourers within a definition of the peasantry, because the landless labourer and the peasant 'are found quite frequently to be one and the same person in different phases of the agricultural cycle', a viewpoint echoed by Cliffe (1977: 202): 'the same individual may have some access to land in his area of origin, defined by [pre-capitalist] customary rights, and may work it along customary patterns, but may also be involved in purely capitalist relations elsewhere by virtue of his buying land or having to sell his labour'. Stavenhagen's analysis (1975) of agrarian societies in Latin America and Africa refers to the development in Mexico of a 'rural semi-proletariat' (p.107) and in Latin America generally of 'the increasing marginalization of the peasantry and the emergence, on a widespread scale hitherto unknown, of a sub-proletariat' (p. 230), while in the Ivory Coast he identifies 'agricultural labourers . . . who . . . do not yet constitute a rural proletariat' (p. 156) but are 'fractions of a class' (p. 158). Elsewhere (p. 92) he states that

> the class situation becomes definitive only when the new position in the production process is stable and permanent. And this is not always the case, because the plantation worker may often leave his work and return to subsistence agriculture in his own village.

One may infer from this that Stavenhagen would not consider such workers as rural proletarians.

These authors, however, seem to beg the question. Many villagers in Papua New Guinea, in common with their counterparts in Latin America and Africa, alternate between self-employment and wage employment. The position taken here is that a villager self-employed on his own land is part of the peasantry but when he is employed by someone else he is a proletarian. It is accepted, nonetheless, that in Stavenhagen's terms Papua New Guinea's rural proletarians do not yet constitute a definitive class. Thus, those who leave their home clans and gain access to land for cultivation elsewhere (by outright purchase, rent, periodic contributions of labour or some other means) are unquestionably part of the peasantry, as are those who have acquired smallholder blocks in formal government-sponsored resettlement schemes. Those who work under agreement for two or more years are part of the rural proletariat while so employed, although if they return to their clans they are likely to resume a peasant or even tribal existence. By this criterion, those employed (but not always for wages: Morauta 1974) on the holdings of big peasants, on indigenously owned plantations purchased under the Plantation Redistribution Scheme, or by cattle owners or smallholders in resettlement schemes, should also be considered rural proletarians, even if they are seasonal workers or are employed on a temporary basis. Those whose employment is full-time and long-term are undeniably rural proletarians.

Rural proletarianisation is perhaps not yet a critical issue in Papua New Guinea but it has important implications for the future. Among these are considerations *vis-à-vis* the goverment's national development goals and strategies which stress the development of rural areas, the spreading of income-earning opportunities through *self*-employment, and the reduction of the present uneven distribution of income (Central Planning Office 1976: 17). For some regions and sectors of society such goals might never have been enunciated. Here it is appropriate to emphasise Marx's warning, cited by Cliffe (1977: 200), that: 'the communal ties of reciprocity in labour use and free access to land [are] unlikely to be effective proof against more individualistic property and labour relations with the growth of commodity production'. In his study of the Kenyan peasantry Lamb 1977: 53) concluded:

> the rapid differentiation of the Kenyan peasantry — and particularly of its most advanced segments — raises the possibility of increasing landlessness, poverty and marginalization for the poorer sections of the peasantry [and] the large-scale development of agrarian capitalism.

Similar trends are apparent in Papua New Guinea. Given the embryonic stage of rural proletarianisation, it is perhaps too early to decide whether a significant rural proletariat will eventually emerge in Papua New Guinea, or whether present trends are simply leading to further differentiation of the peasantry. Referring to the fragmentation and limited class solidarity of the French peasantry in the nineteenth century, Marx characterised it as 'a sack of potatoes'. By the same token, the twentieth century peasantry of Papua New Guinea may prove to be a 'sack of *kaukau*'.

References

Amarshi, A. and Blaxter, L. 1975. Notes towards the definition and understanding of the 'working class' in Papua New Guinea. (Paper read at the Conference on Studies in Class Analysis, August 1975, Sydney.)

Central Planning Office 1976. *The Post-Independence National Development Strategy*. (Papua New Guinea Government White Paper.) Central Planning Office, Waigani, Port Moresby.

Cliffe, L. 1977. Rural class formation in East Africa. *Journal of Peasant Studies* 4(2): 195-224.

Connell, J. 1977. The economics of big-men: reflections on a Siwai paradigm. (Work-in-progress paper, Department of Economics, Research School of Pacific Studies, The Australian National University, Canberra.)

Crocombe, R. G. 1965. *The M'Buke Co-operative Plantation*. New Guinea Research Unit Bulletin No. 7, The Australian National University, Canberra.

Densley, D.R.J. 1972. Rural industries and rural wages in Papua New Guinea. Pp. 322-42 in M.W. Ward (ed.), *Change and Development in Rural Melanesia*. Research School of Pacific Studies, The Australian National University, Canberra and University of Papua New Guinea, Port Moresby.

Fingleton, J.S. 1977. Unpublished background paper for review of Plantation Redistribution Scheme. Department of Lands and Environment, Port Moresby.

Fingleton, J.S. in press. Land policy in Papua New Guinea 1972-1977. In J.A. Ballard (ed.), *Policy-making in a New State*. University of Queensland Press, Brisbane.

Finney, B.R. 1973. *Big Men and Business: Entrepreneurship and Economic Growth in the New Guinea Highlands*. Australian National University Press, Canberra.

Franklin, S.H. 1962. Reflections on the peasantry. *Pacific Viewpoint* 3: 1-26.

Gerritsen, R. 1975. Aspects of the political evolution of rural Papua New Guinea: towards a political economy of the terminal peasantry. (Unpublished seminar paper, Department of Political Science, Research School of Social Sciences, The Australian National University, Canberra.)

Gilbert, A. 1974. *Latin American Development: A Geographical Perspective*. Penguin, Harmondsworth.

Good, K. 1976. The formation of the peasantry in Papua New Guinea. (Unpublished seminar paper, Department of Political Studies, University of Papua New Guinea, Port Moresby.)

Heaney, W.H. 1977. *Preliminary Findings and Implications of a Study of Rural Migration and Social Change in the Wahgi Valley, Western Highlands Province*. (A report to the Western Highlands Provincial Government, Papua New Guinea.) East-West Center, Honolulu.

Howlett. D. 1962. A decade of change in the Goroka Valley, New Guinea: land use and development in the 1950s. Unpublished Ph.D. thesis, The Australian National University, Canberra.

Howlett, D. 1973. Terminal development: from tribalism to peasantry. Pp. 249-73 in H.C. Brookfield (ed.), *The Pacific in Transition: Geographical Perspectives on Adaptation and Change*. Arnold, London.

Howlett. D. *et al.* 1976. *Chimbu: Issues in Development*. Development Studies Centre Monograph No. 4, The Australian National University, Canberra.

Joel, A. 1974. Plantation labour survey into the Highlands Labour Scheme. (Unpublished report for Department of Labour, Commerce and Industry, Papua New Guinea.)

Lamb, G. 1977. The neocolonial integration of Kenyan peasants. *Development and Change* 8: 45-59.

Lipton, M. 1977. *Why Poor People Stay Poor: A Study of Urban Bias in World Development*. Temple Smith, London.

Low, D.A. 1976. The Asian revolutions of the mid-twentieth century: an Australian perspective. (Paper read at First National Conference of the Asian·Studies Association of Australia, Melbourne.) Melbourne (mimeo).

McGregor, A.M. 1971. *Langandrewa and M'Buke, Corporate Indigenous Plantations*. New Guinea Research Bulletin No. 43, The Australian National University, Canberra.

McKillop, R.F. n.d. [1975]. Problems of access: agricultural extension in the Eastern Highlands of Papua New Guinea. Department of Agriculture, Port Moresby (mimeo).

McKillop, R.F. in press. Plantations: a relic of the past? In *Yagl-Ambu* 5 (1979).

May, R.J. and Skeldon, R. 1975. *Internal Migration in Papua New Guinea: An Introduction to Its Description and Analysis*. Discussion Paper No. 4, New Guinea Research Unit, Boroko.

Mintz, S.W. 1973. A note on the definition of peasantries. *Journal of Peasant Studies* 1(1): 91-106.

Mintz, S.W. 1974. The rural proletariat and the problem of rural proletarian consciousness. *Journal of Peasant Studies* 1(3): 291-325.

Morauta, L. 1974. *Beyond the Village: Local Politics in Madang, Papua New Guinea.* London School of Economics Monographs on Social Anthropology No. 49. Athlone Press, London.

Mortimer, R. 1975. *Social Science and the Peasant: A Case of Academic Genocide: Public Lecture by Rex Mortimer Professor of Political Studies.* University of Papua New Guinea, Port Moresby.

Ploeg, A. 1975. *Spontaneous Rural Resettlement In and Near the Chimbu District in 1972-73.* Discussion Paper No. 5, New Guinea Research Unit, Boroko.

Rowley, C.D. 1958. Labour administration in Papua and New Guinea. *South Pacific* 9(11): 540-8.

Shanin, T. 1971. Peasantry: delineation of sociological concept and a field of study. *Archives Européennes de Sociologie* 12: 289-300.

Shanin, T. 1973. The nature and logic of the peasant economy — 1: a generalisation. *Journal of Peasant Studies* 1(1): 63-80.

Shanin, T. 1977. Measuring peasant capitalism. *Economic and Political Weekly* 12: 1939-47.

Smith, D.M. 1977. *Human Geography: A Welfare Approach.* Arnold, London.

Smith, D.W. 1975. *Labour and the Law in Papua New Guinea.* Development Studies Centre Monograph No. 1, The Australian National University, Canberra.

Spate, O.H.K. 1973. Foreword. Pp. v-vii in H.C. Brookfield (ed.), *The Pacific in Transition: Geographical Perspectives on Adaptation and Change.* Arnold, London.

Stavenhagen, R. 1975. *Social Classes in Agrarian Societies.* Anchor Books, New York.

Tomassetti, W.E. 1966. Community development and the Chimbu. Pp. 85-94 in E.H. Hipsley (ed.), *An Integrated Approach to Nutrition and Society: The Case of the Chimbu.* New Guinea Research Unit Bulletin No. 9, The Australian National University, Canberra.

Territory of Papua and New Guinea 1966. *Report of the Board of Inquiry Investigating Rural Wages and Related Matters.* Department of Labour, Konedobu.

Territory of Papua and New Guinea 1970a. *Report of a Board of Inquiry Investigating Rural Minimum Wages under the Native Employment Ordinance 1958-1968, Minimum Wage-Fixing Machinery and Related Matters* [the Cochrane Report]. Government Printer, Port Moresby.

Territory of Papua and New Guinea 1970b. *The Structure of Unskilled Wages in Papua and New Guinea* [the Isaac Report]. Goverment Printer, Port Moresby.

von Fleckenstein, F.W. 1974. An analysis of the tobacco industry in Papua New Guinea. (Unpublished seminar paper, New Guinea Research Unit, Port Moresby.)

Young, E.A. 1977. Simbu and New Ireland migration. Unpublished Ph.D. thesis, The Australian National University, Canberra.

Chapter 11

Tourism in Papua New Guinea

The Last Resort

*D. A. M. Lea**

An efficient, modern tourist industry, often sought by businessmen and politicians in Third World countries (Department of Labour, Commerce and Industry 1976: 37), may well be antithetical to development that is truly concerned with 'raising levels of life sustenance, human dignity and freedom' (Goulet 1971: 87), and which 'can be assessed only in terms of the total human needs, values and standards of the good life and the good society perceived by the very societies undergoing change' (Goulet 1971: 85).

Tourism certainly generates economic advantages for some but it does not help the development of the majority. While the benefits appear obvious, it is too easy to forget that people in the tourist industry are skilled in the art of salesmanship and promotion. The industry's basic objective is to convince both hosts and potential tourists that expensive, luxurious and often exhausting travel is a worthwhile experience for those who can afford it and good business for the hosts. A measure of the industry's success is that tourism is big business, it is growing, and it is the largest single item of world trade. Further a conventional wisdom holds that tourism can create jobs, has spectacular multiplier effects, can earn badly needed foreign

* *David Lea is Professor and Chairman of the Department of Geography, University of New England. After graduating from Adelaide University, he taught in a secondary school for three years and then undertook his doctoral degree in the Department of Geography, Research School of Pacific Studies, The Australian National University, 1961-4. For five years he taught at The University of Papua New Guinea.*

exchange, diversifies and stabilises the economy, and can create economic opportunities in rural areas. In addition to these economic virtues some say that tourism promotes interaction between people of different cultures, helps countries to discover their national identities or even provides a reason for, and method of, retaining traditional crafts and ceremonies (Peat *et al.* 1973: 3).

All these claims can be questioned in developed countries where tourism develops a culture and an environment of its own. Sites are so altered that they lose their original unique and pleasing qualities and few economic advantages accrue to most of the original occupiers. In developing countries the conventional wisdom needs to be looked at most carefully because international concrete and carpeted hotels, cruise ships, air-conditioned buses, packaged tours and promises of sun, sex, exotic places, quaint and friendly people have little to do with the realities of poor people living in poor countries. In Papua New Guinea (Fig. 11.1) problems arising from its recent contact with the West, great linguistic and cultural diversity, poor access, alternative developmental needs and priorities, inadequate infrastructure for all types of tourism and its dependency on the

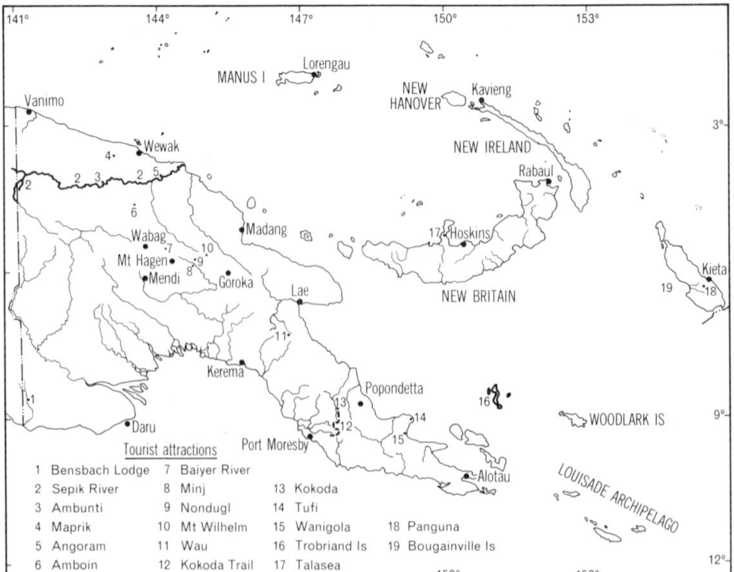

Figure 11.1 Papua New Guinea. Places mentioned in Chapter 11.

economies of the capitalist world accentuate these difficulties.

Informed evaluation of tourism in Papua New Guinea is difficult because data are not available. The Draft National Tourism Plan of 1977 (which must contain masses of data) had not been released at the time of writing (August 1978), and most of the really important human impacts — especially the long-term ones — are impossible to assess and to measure with any reliability. Further, any discussion of tourism inevitably involves questions of opinion, prejudice and vested interest and it is relatively easy to present tourism as either a promising and lucrative industry or as a convenient scapegoat for many of the problems caused by social and economic change. Neither view is correct. Tourism is an outward and visible manifestation of much that is exploitative, intrusive and garish in Western culture but its total impact is probably slight compared with urbanisation, industrialisation, education, television, newspapers, political ideas and the operations of transnational corporations in other industries. However, there is conflict between the reality of international tourism and many of the Eight National Aims espoused in the National Development Strategy (Central Planning Office 1976). Among other things these 'Aims' call for an increase in the proportion of the economy controlled by nationals, decentralisation, self-reliance, locally raised revenue and 'necessary' government control and involvement. Conventional tourism is incompatible with most of these aims because it is enclave or town orientated and it is dependent on international companies which, for their own financial gain, recruit, direct and control the movement of most tourists and which offer 'crumbs' to local people and local governments (Samy 1973).

In spite of this it seems unrealistic to suggest that Papua New Guinea should not have a tourist industry. Without a revolutionary restructuring of society no poor or dependent country can afford to reject any income-earning opportunity or to close its borders to Western capitalists while seeking Western capital. It is also likely that the leaders in Papua New Guinea perceive a need to demonstrate to the rest of the world that their nation is modern and civilised so they readily approve prestige buildings in the main towns, an international airline and an 'acceptable' welcome for rich overseas visitors. Also Papua New Guineans like to travel both to see places and to visit friends. Thus a tourism policy is necessary to relate realities with national needs and aspirations and to allow for effective planning.

The Tourist Industry in Papua New Guinea

Until the 1960s tourism was not encouraged for it was feared that tourists would have a detrimental effect on the indigene and large areas of the country were not under effective control (International Bank for Reconstruction and Development 1965: 219). In 1961, for example, only 9,648 short-term visitors entered the country. Of these 2,459 were businessmen or short-term workers, 1,486 were in transit and about 3,000 were schoolchildren visiting parents and residents returning for short-term visits. Thus in 1961 there were only about 2,720 'holiday' visitors and probably over half of these were friends and relatives.

As a result of the World Bank visit in 1963 the Australian Administration adopted a more positive attitude to tourism and in 1966 set up a statutory body, the Papua New Guinea Tourist Board, whose task it was to promote tourism. 'Apart from the production of many bright promotional posters and the setting up of advisory bodies involving private sector tour operators, there is little to suggest that the Board had much impact' (Williamson 1977: 31). Nevertheless visitors gradually increased in number until in 1975 the Board was replaced by the Office of Tourism, a division within the Department of Labour, Commerce and Industry.

In 1976 there were 14,493 'holiday' visitors to the country but this reflects a persistent decline in the number of tourists since 1973 when there were 21,925 (Fig. 11.2). Neither the world-wide recession of the tourist industry in the first half of the 1970s nor the fact that fewer friends and relatives (estimated by Gibbings (1973: 10) to be 40 per cent of all tourists in 1973) are visiting the diminishing European population accounts for this decline. It may indicate that the industry is reluctant to send tourists to a place without either a 'white welcome' or an established tourist infrastructure. It may be because Trans-Australia Airlines and Ansett Airlines of Australia no longer operate in Papua New Guinea and therefore do not bother to 'sell' Papua New Guinea on the Australian market. It may be because the potential tourist thinks self-government has resulted in excessive political instability and 'risk' for travellers. It may be that travel in Papua New Guinea has become more expensive than travel in other tourist areas.

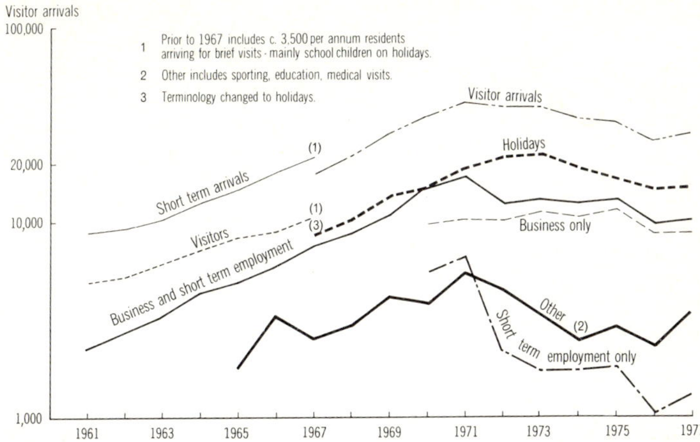

Figure 11.2 Visitor arrivals in Papua New Guinea 1961-77. (*Source:* Bureau of Statistics Papua New Guinea, Bulletins relating to overseas and international migration.)

The majority of short-term visitors are holiday-makers. Figure 11.2 shows that consistently about half of all short-term visitors are on holiday and about a third are on business. The category 'short-term employment' shows the greatest decline and this follows the completion of the construction phase of Bougainville Copper Ltd and the implementation of government controls on people entering the country for employment purposes.

In 1976 64 per cent of all visitors were from Australia but other important source countries were Canada (8 per cent), United Kingdom (7 per cent), Japan (5 per cent) and New Zealand (4 per cent). The dominance of the Australian market is shown by the coincidence of three peaks in visitors with the school holidays in Australia (Fig. 11.3). The August peak is the greatest, and this is also the time of the Highlands Show in either Goroka or Mount Hagen.

The holiday-maker does not stay long in Papua New Guinea. A survey by Gibbings (1973: 17) indicated that the average holiday-maker stayed only 12.2 days and all categories of short-term visitors (excluding itinerant workers) stayed on average 18.7 days. A more recent tourist expenditure survey carried out by the Office of Tourism in 1976 found that the average holiday-maker stayed for only six days and all categories of short-term visitors stayed only 8.1 days. Gibbings (1973: 83) estimated that visitors spent an average of

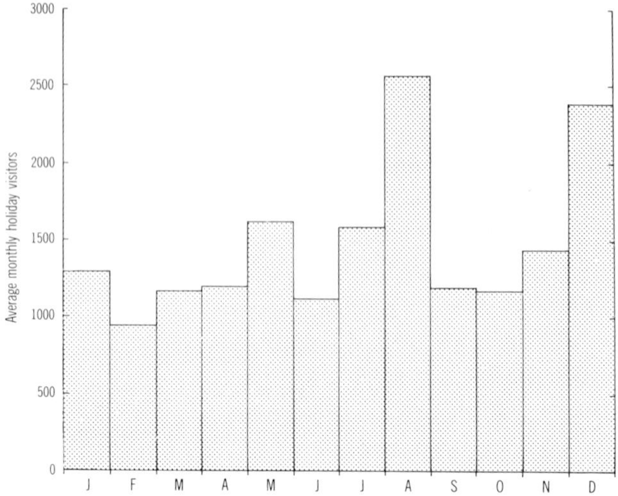

Figure 11.3 Average monthly holiday visitors arriving in Papua New Guinea 1969-77. (*Source:* as for Figure 11.2.)

$A246 within Papua New Guinea and those on holidays spent $A376. The Office of Tourism survey found an average visitor expenditure of $A373. Allowing for inflation and some underestimation it is likely that the average tourist in 1976 spent $A400 so that the 28,729 visitors generated nearly $A11.5 million, a figure that is higher than the K5.6 ($A5.6) million of direct purchases in the domestic market by 'non-resident households' in 1976 (Bureau of Statistics 1978) and much lower than the estimated expenditure by overseas visitors of some $A14.1 million in 1970 by Parker (1973: 44). The range of estimates indicates mainly the unreliability of the figures but this has not discouraged some protagonists of the tourist industry (Peat *et al.* 1973: 13) from comparing these figures with the value of principal commodities exported (see Table 11.1). This comparison is dubious and will be discussed later.

The final characteristic of present-day tourism in Papua New Guinea is that it is expensive. The first director of the Office of Tourism reportedly stated (*Post Courier* 18 May 1976) that hotel accommodation costing $A30 in Port Moresby would cost $A22 in Hawaii and $A15 in Manila. Compared with similar accommodation in New South Wales, hotels in Papua New Guinea are about 20 per

Table 11.1. Value of principal commodities exported from Papua New Guinea 1977*

Commodity	Kina (000)	$A (000)
Copper ores and concentrates	201,050	228,466
Coffee	143,441	163,001
Cocoa	86,946	98,802
Timber and timber products	23,971	27,240
Fish	19,158	21,770
Copra and copra products	36,745	41,756
Palm oil	8,965	10,188
Tea	9,765	11,097
Rubber	2,896	3,291
Other	6,899	7,839
Total (all domestic produce)	539,836	613,450

* The kina had parity with the Australian dollar until mid-1976. In June 1977 $A1 = K0.88 and in August 1978 $A1 = K0.80.
Source: Bureau of Statistics Papua New Guinea.

cent dearer; air fares in Papua New Guinea are approximately 17 per cent higher per route kilometre than those on Australian internal routes (and 97 per cent higher than those in Malaysia); and a thirteen-day packaged tour from Sydney costs $A1,450 in Papua New Guinea compared with similar or longer tours costing $A960, $A820 and $A1,260 in Bali, Fiji and Malaysia advertised by Qantas Airways Ltd. Even a nineteen-day off-season tour in Europe costs only $A750.

The rest of this essay is divided into two parts, the first of which questions the conventional wisdom about tourism and shows that there are, or may be, conflicts with developmental strategies, while the second suggests a *modus operandi* for the industry in Papua New Guinea. The basic theme is that the government should put little money into the industry and tourism should be low in all investment priorities. In an attempt to make tourism compatible with, or at least not in excessive conflict with, the Eight Aims, conventional tourism should be isolated (also suggested by President Nyerere of Tanzania (Bryden 1973: 2)) and a new appropriate form of tourism developed for the country. If the industry cannot accommodate itself to these changes, Papua New Guinea may well be better off with a tourism policy that discourages or nationalises the industry. The latter option means unwarranted expenditure, setting up agencies in the tourist source regions, and competing with the international tourist businesses — an unrealistic option for a poor country.

> In 1971 Wolfers (1971: 4) wrote that there is a tendency among investigating experts and consultants from the large international agencies, to forgo their usual statistical analyses, painstakingly charted graphs, and carefully compiled tables, when investigating the tourist potential of the South Pacific, for ecstatic poetry and projected whimsy.

This is still true although, fortunately, some government economists, politicians and concerned laymen are asking important questions. It is also significant that the Draft National Tourist Plan still has not been released even though it was completed in July 1977 and its preparation was the first priority task of the Office of Tourism. From reports leaked to the *Post Courier* (the national daily) it would seem that the Plan espoused conventional wisdom; thus the alleged economic advantages and then the alleged socio-cultural advantages of this conventional wisdom should be examined.

Economic advantages

While it is true that tourism diversifies the economy, increases national income and earns foreign exchange, many of the benefits do not go to nationals or to the Papua New Guinea Government. The tourist industry needs many imports: for example, the flag-carrying airline, Air Niugini, has to buy planes, spare parts, fuel and many of the high-level management and technical staff from rich countries, and Air Niugini is partly owned by foreigners (16.7 per cent of the shares held being by Ansett Airlines of Australia). The average, and most certainly the wealthy, tourist normally drinks the beverages and eats the foods that he knows and likes, and seeks accommodation and services to which he is accustomed. Thus much of the money that a tourist brings into countries such as Papua New Guinea is immediately lost by overseas promotion, importing goods, and exporting capital to buy the equipment and materials to build up a network of hotels, air-conditioned buses and other facilities which may not be appropriate to the development of other sectors of the economy. Additionally, interest payments on borrowed capital as well as remitted components of expatriate salaries, dividends and profits mean that there is probably a much higher import leakage than the 'direct' import leakage of 23.8 per cent calculated by Gibbings (1973: 91).

Considering the high proportion of the industry in the hands of foreigners (see later) and the high rate of return on capital demanded by overseas interests (which accounts for the high prices), 'total' import leakage is probably around 40 per cent. In this respect Britton's findings in Fiji are of interest. He observed (personal communication) that tourism has a greater multiplier effect on imports than all the other major key sectors of Fiji and it therefore has a high propensity to increase national import requirements. The Central Planning Office of Fiji (1975: 169) found that the multiplier effect on Gross National Product (at 0.71) was lower than all other sectors except mining and quarrying.

The two estimates (Peat *et al.* 1973: 14; Gibbings 1973: 104) of the multiplier effect in Papua New Guinea are both based on the work of Parker (1973). Their estimates of a multiplier of about 0.8 seem basically acceptable although they may make inadequate allowance for the way the expatriate population remits income and enterprise surpluses and they may have underestimated the value of imports needed to support the industry in Papua New Guinea. While this figure is lower than those usually quoted by the industry (e.g. Burkart and Medlik 1974: 60; Robinson 1976: 127), advocates of the tourist industry still manage to misuse this figure. They say correctly that each $100 of tourist expenditure results directly or indirectly in local production of $80 (and indirect taxes of $10 and imports of $68) but they do not say what the multiplier effects are in other sectors or industries; they may well be higher. Multipliers are usually calculated at a national level and they ignore the fact that rural areas import from both overseas and main urban centres so the multiplier effect in those rural and isolated areas that need local production would be minimal and tourism would certainly accentuate the 'urban bias' described by Lipton (1977). As Bryden (1973: 5) argues, 'the multiplier cannot be used as a reliable indicator of the benefits to be derived from tourism except under very restrictive assumptions which are unlikely to be met in the real world'.

It is extremely difficult to estimate the economic benefits that tourism brings to nationals. Using tourist expenditure impact analysis Harcourt (1977) estimates that, of the domestic expenditure of overseas tourists in Papua New Guinea during 1976, nationals received 27.3 per cent of the tourist expenditure in profits and wages whereas foreigners received 33.6 per cent (Table 11.2). Thus with resident and non-resident foreigners receiving a small part of government revenue by direct (e.g. salaries and consultancies) and

Table 11.2. Impact of international holiday tourist expenditure in Papua New Guinea 1976

		(per cent)
Nationals:	Profit	8.8
	Wages	18.5
Foreigners:	Profit	11.5
	Wages	22.1
Government revenue		15.6
Imports		23.6
Total		100.1

Source: Harcourt 1977: 181.

indirect (e.g. government services and subsidies) means, all of that committed to imports, and most of that spent on international travel to the country, about 60 per cent of the tourist dollar put aside for 'travel to and in Papua New Guinea' does not accrue to nationals or the government.

Without detailed research over a long period it would be difficult to ascertain trends. An impressionistic view is that local ownership is becoming more common. For example, the Bird of Paradise Hotel in Goroka is owned by the Eastern Highlands Provincial Government and other provincial hotels are owned by local development corporations or groups of local businessmen. However, even if it is assumed that this trend is widespread, most tourists stay in hotels owned by foreigners (e.g. Davara Hotels, Steamships Hotels, Kamsta Hotels and Travelodge Hotels) and some local ownership could involve a good deal of 'dummying' by the indigenous wives of foreigners and foreign ownership of local companies.

Much is made of the fact that tourism creates jobs. While this is true they tend to be lowly-paid occupations (some forms of prostitution excepted) and tourism has seasonal peaks which result in marked fluctuations in the level of employment. Williamson (1977: 37) estimates 'that the enumerated wage and salary earning indigenous work force directly employed in the industry is not much more than one thousand out of a total enumerated work force in June 1973 of 118,482, or in other words, less than one percent'. Add some artefact makers, guides, beggars and prostitutes from the informal sector and the total figure would not be much more impressive. Regardless of the numbers involved it is more significant that the labour-capital ratio is low for tourism and that alternatives in the service, industrial or agricultural sectors may offer better opportunities. For example, in Tunisia between 1956 and 1971 the cost of creating a job in the hotel industry was between $A13,300 and $A20,300 while in the manufacturing industry it was $A12,700 (Bugnicourt 1977a: 1). Further, even though tourism and associated industries create some jobs that do not need much training, most positions in fact require people with some linguistic abilities and an understanding of the values of people from other countries. Thus even the lowly positions may divert skills from industries with higher priorities.

Tourism has three undesirable economic impacts that are difficult to quantify. First, the industry is very dependent on economic conditions in the affluent capitalist world. The capital to establish the infrastructure, many of the goods to cater for the needs of tourists, and the tourists themselves all come from overseas. Second, tourism leads to spatial inequalities because it is corridor or enclave orientated. Britton (personal communication) has shown that in Fiji 83 per cent of gross industry turnover from tourism is generated in eleven towns of which four (Suva, Nadi, Sigatoka and Lautoka) are responsible for 81 per cent. He points out that nearly all tourist expenditure takes place in areas that were historically part of Fiji's colonial export economy or in resort complexes which transfer income to company headquarters and other businesses in the three main towns of Suva, Nadi and Lautoka. It is conceivable that a similar pattern applies to Papua New Guinea, with most tourist expenditure taking place along a Port Moresby-Lae-Mount Hagen axis as the more isolated resort centres such as the Karawari Lodge at Amboin, the Ramada Lodge at Minj and the Bensbach Wildlife Lodge presumably acquire much of their capital and provisions from urban centres. It may be hypothesised that, with increasing isolation and decreasing size of town or resort, less tourist expenditure filters through to local people. Third, scarce resources often may go to provide services and facilities for wealthy tourists so that schools, hospitals, feeder roads are neither built nor upgraded (Ala'ilima 1977). The expensive and inappropriate infrastructure, such as international hotels and casinos (*Post Courier* 18 October 1977), may well divert government spending away from cheaper and yet adequate items that would serve the Papua New Guinean society at large. The setting up of visitors' bureaux and government offices and the provision of services for tourists mean that 'less privileged tax payers are subsidising holidays of the privileged' (Bugnicourt 1977a: 1). For example, the Office of Tourism cost the Papua New Guinea taxpayer K348,000 ($A395,000) in 1976-7. Apart from the preparation of the Draft National Tourist Plan, the main function of this Office is to promote Papua New Guinea as a tourist destination: one result of this is to increase the profits of the tour operators because it saves them the cost of promotion (Williamson 1977: 40).

Most of this section has related to short-term benefits and costs. The long-term impacts are even harder to evaluate. Will tourism lead to the development of a comprador class? Will it lead to increasing dependency? Will economic and spatial structures be established which do not serve the needs of the developing society (Lea 1975)? And, finally, what are the long-term alternatives? Papua New Guinea is fortunate in that it has a wide resource base, internal markets and good access to Australian and Japanese markets. Unlike the situation in some small Pacific islands, agriculture, mining and manufacturing may be reasonable alternatives.

Socio-cultural advantages

It is often said that tourism promotes better international understanding (Burkart and Medlik 1974: 57). However, in towns and villages tourists rarely talk to local people because they cannot speak local languages or *lingue franche,* and even if local people speak English they have nothing in common to talk about except the prices of commodities offered for sale or trite and condescending topics. The tourist may say that he likes meeting people but it is a rare one that actually does. The actual extent of most tourist interaction is to be photographed beside an 'arse-grassed' local or to photograph people as if they were in a zoo or while they perform pseudo-cultural and packaged events. Confronted with beggars, poor people, dirty people or primitive people, most tourists feel threatened and are happy enough to exchange smiles and waves from a bus and interpret the courteous or mocking gestures as 'aloha'. In spite of the rhetoric: 'Most tourists are more interested in looking at scenery and famous monuments than in entering into contact with the living culture: they respond merely to their own expectations' (Bugnicourt 1977b). In this respect tourists may actually work against better international understanding for they return to their own countries as ill-informed 'experts' (e.g. Papua New Guineans are 'happy', 'lazy', 'stupid' or 'friendly'), and probably use the tourism experience to reinforce their own initial prejudices. Also their portly figures and artefacts of affluent living give host people the wrong ideas about the cultures and problems of the countries from which the tourists come as well as unrealisable aspirations. Bugnicourt (1977b) says tourists may well 'stir the fancy of poverty-stricken masses'. Some local people maintain that tourists create envy and resentment (e.g. Talyaga 1974) or that they represent a new form of cultural imperialism. V. S. Naipaul, the West Indian novelist, put this point of view graphically when he wrote

> Once, because of their wealth a people had been enslaved; now because of their beauty, a people are being dispossessed . . . Every poor country accepts tourism as an unavoidable degradation. None has gone as far as some of these West Indian islands, which, in the name of tourism, are selling themselves into a new slavery. (Naipaul 1969: 210.)

To claim that tourism provides an opportunity to preserve cultures is fatuous (Peat *et al.* 1973: 3) for, as tradition and custom become commercialised, they must undergo stress at the best or be demoralised, dehumanised or degraded at the worst. As Fraser (1972: 6) says, 'When Bali dances for the gods become tourist attractions, and African war dances become money spinners, their whole character and authenticity gets altered'. Beier (1977) talks about tourism creating a 'pseudo-culture' which bears only superficial resemblance to the authentic culture. Writing of Tolai art, Simet (1976: 1) argues that it 'is made up of many interlocking elements' and 'The most important element of any art form is the setting'. He suggests that it is not right for anybody to help disintegrate art forms and ceremonies in order to please and entertain people who do not appreciate them fully. Simet is concerned with the impact on the host societies of strangers concentrating on a fragment of culture they call art but which also has social, economic and spiritual functions. Certainly the profanation of places and of objects of worship, the commercialisation of culture, the perversion of living and developing ceremonies and the dissociation of time and place must affect cultures that are brittle and subject to many other great stresses. What do the mudmen of the Asaro Valley think of their unique ceremony when they put on, at demand, a mediocre 3½ minute performance, that takes only 25 minutes to prepare, as often as seven times a week for K5 a performer? What respect do young men and women have for their traditional dances when they perform them in a carpeted fantasy-land in a way that is completely out of context? It is a form of prostitution:

> Both mother and daughter,
> Working for the Yankee dollar.

Of course cultures change and their members have to adapt to new circumstances, but tourism often encourages an excessively rapid rate of change and it is a form of change that local people cannot control. A new internationalism demands that tourists arriving in any Pacific Island country should be met with Hawaiian leis, hula dancers and kava ceremonies or, as Spate (1965) satirically suggests, foreign dancers could be imported to perform 'traditional' dances. Perhaps 'culture' should only be served to tourists when performances and exhibitions are set up by local people in their own way and on their own terms.

Finally tourism has both a cultural and environmental impact which destroys or lessens the essential attraction. Thus the Gold Coast, Niagara Falls and the Asaro Valley are to be avoided by the lover of beautiful beaches, the honeymooner wanting a quiet place and the person really interested in Papua New Guinean cultures. As many of the best tourist environments are beautiful and fragile ecosystems, is it possible that the industry has fouled its own nest in the developed world and is now rushing to consume the environments of developing countries?

Some Possibilities for Papua New Guinea

Overall there seem to be few economic, social or cultural advantages from tourism for Papua New Guinea. Yet, for reasons already pointed out, a tourism policy is necessary. To help resolve the contradictions three compromises may be suitable for Papua New Guinea.

The option for the rich international tourist

There is no doubt that Papua New Guinea has an appeal for a very small minority of wealthy tourists who have special interests such as travel in unusual places, buying artefacts, bird watching or collecting orchids. These tourists should be catered for by having international standard hotels in five to seven towns along a 'main tourist route' which could link Port Moresby, Lae, Goroka, Mount Hagen, Madang and Rabaul. Highly efficient and narrowly channellised excursions or spurs could lead off this main route to resorts or selected tourist attractions. These few resorts would all be in sparsely populated areas such as Tufi, the Madang coast, Baiyer River, Talasea, some isolated coral island and the lodges already established. The government should not put any resources into these resorts and any further developments by private enterprise should be under licence from the government and with such safeguards as those established by the Fijian Land Trust Board (Nayacakalou 1972). These safeguards include local equity, representation on the boards of directors, rental based on unimproved capital value or gross receipts (whichever provides the greater returns), inspection of site plans before the lease is signed and regular review of leases.

It should be policy that these rich tourists would make little contact with local people. They would move around, hygienically sealed in air-conditioned packages, and they should pay well for all services. This sort of tourism should offer investors a rapid return on investment and it may attract potential investors and simultaneously cater for businessmen, overseas experts and advisers visiting the main towns.

Many of the norms of conventional tourism would apply to this type of tourism. The government would have to accept that tourists be directed to towns and enclaves so that the industry can be controlled and supervised and indigenous people protected from the economic and cultural dangers. The compensation would be that the government would not need to contribute anything for promotion or infrastructure costs (assuming most investment in resorts and tours is by private enterprise) and government officials would only supervise the investment guidelines and watch for land, social and environmental problems.

The option for the relatively poor Australian tourist

Australia is at present the source of most tourists. However, Papua New Guinea has the potential to develop a form of tourism that is less dependent on overseas capital, tour operators and international airlines, and that can compete with the present places which attract less affluent Australian overseas tourists. This group is represented by tertiary level students and the mobile, young middle class who might be prepared to spend $A500 to $A800. About $A300 would be absorbed by air fares to Papua New Guinea and a further $A100 at least on internal air or sea travel, with the balance being spent on accommodation, food and souvenirs. This type of tourism could be planned in such a way that it would not disrupt local people, would not require an elaborate and expensive infrastructure, and would give the bulk of any financial rewards to people in rural areas. These tourists should be encouraged to travel along designated routes, using hostels or cheap guest-houses in towns and rest-houses near villages. They would have to use local transport facilities such as public buses and public motor vehicles and each hostel or rest-house need only provide beds, mattresses, showers, toilets, cooking facilities and mosquito nets.

Apparently the National Tourist Plan (according to the *Post Courier* 24 August 1977) suggested a village stay scheme but this could disrupt village life and usually would be an unwelcome intrusion (many of us have seen how a single anthropologist or geographer who is trained not to disrupt village life often does not succeed). Rest-houses should be built and operated by villagers under licence from councils or provincial governments. They should not be in villages but along roads at least three kilometres from the nearest village and preferably midway between villages. Each rest-house, built of local material like the old *haus kiaps,* would have a resident superintendent to collect fees, to guard against theft and damage, and to report delinquents to the police. All such travellers could have a registration card which would be surrendered when checking in, and returned when the bill is paid and the place tidied. The superintendent could be assisted by a full-time or part-time cleaner/gardener. Any traveller using these rest-houses would agree to observe a basic set of rules which could be modelled on those of the Youth Hostels Association of the United Kingdom. All rest-houses would be registered and the main task of government would be to approve location sites, to inspect them and to attempt to get them along routes serviced by existing infrastructure.

Advantages of this type of tourism are that tourists spend money in rural areas and it accrues to local people. Infrastructure costs would be very low. The problems are that there has to be some sort of national organisation to ensure that linkages and routes are provided for the tourists, and national and provincial governments would have to ensure that villages are not over-exposed to tourists and that both tourists and hosts are properly controlled. It would be impossible to stop villages or councils setting up rest-houses on their own but all registered rest-houses would have to be part of a route or a system. This means that the benefits and problems of this sort of tourism could not be spread throughout all rural areas. As pointed out earlier this is unavoidable because tourism is route and enclave orientated and it is desirable to keep tourism in those places that are either already subject to rapid social change (i.e. the towns and along the main roads) or in very sparsely populated areas.

The indigenous tourist

The third and final type of tourist development is that which caters for the local traveller. Judging by the high degree of migration (May and Skeldon 1977) and visiting (Young 1973), travel is a popular activity in Papua New Guinea and may have, at least to some extent, replaced initiations as a means of proving adulthood. Initially any developments would have to augment the *wantok* system (where travellers or migrants are dependent on others from the same area or group of origin) and provide basic shelters, toilets, washing and cooking facilities in each town, for say one or two kina a night. Larger towns could have pensions, lasmens or guest-houses which could be locally owned and operated, and it is interesting to see the development of a few such places already. Certainly tourist planning should cater for the needs of indigenous as well as overseas travellers for it offers excellent opportunities for local entrepreneurship and may ultimately lead to accommodation which is appropriate for budget travellers and mobile bureaucrats.

Safeguards

To guard against the worst features of conventional tourism and to provide policy guidelines, some safeguards which go further than the platitudes of the National Investment Strategy (Department of Labour, Commerce and Industry 1976: 37-8) would have to be designed in consultation with the industry. Assuming that state ownership of tourism is unacceptable and unwise at the moment, the private sector would have to accept limitations and new approaches to the industry. These in themselves would probably stop rapid growth of the industry.

Some safeguards could include the following:

1. Legislation to ensure a reasonable amount of local equity or profit rental to landholders.
2. Local representation on boards.
3. Provisions that the private sector of the industry should train local personnel, and that indigenous guides should accompany each group to act as interpreters, educators and advisers to tourists. At the same time they would watch for behaviour which may be demoralising or degrading to local people or local custom.
4. Provisions that research should be carried out under the auspices of local research groups such as the University of Papua New Guinea or the Institute of Applied Social and Economic Research into those poorly researched problems mentioned in the first part of this essay.
5. Provisions for compulsory social and environmental impact statements for any proposed development by individuals, firms or institutes *not* involved in the tourist industry.
6. Preparation of a register of all scenic attractions, overnight accommodation and modes of transport at present available to the tourists.
7. The drawing up of a code of ethics for travellers which should be given to all tourists when they enter the country (Lea 1977).
8. Provisions of incentives for businessmen to use local produce and local skills while increasing tariffs on imported goods or materials which could be grown or made in the country.
9. A fair and reasonable tax on tourists so that the Papua New Guinea government is not subsidising the overseas tourist. (Such a tax would help pay for air navigation charges, custom services, infrastructure costs and any promotion carried out by the government: it should also take into account spin-off benefits and could be reduced if more benefits accrue to Papua New Guinean nationals.)

Conclusion

Tourism, like other aspects of modernisation, may appear to create developmental opportunities but in fact it probably encourages dependency, spatial and economic inequalities, a net flow of resources from 'developing' to 'developed' countries (Bryden 1973: 60-8), and a retreat from traditional values. Without revolutionary change it is as hopeless for a country like Papua New Guinea to stop internal and international movement of people as it is to stop urbanisation and other forms of modernisation: thus every attempt should be made to control movements and make them serve the national interest. Because tourism by its very nature conflicts with the idealism of the Eight National Aims of the Papua New Guinea government the choices are cruel and somewhat limited. First, the Eight Aims can be ignored and private enterprise can develop the industry along conventional lines; second, tourism can be discouraged officially or even forbidden but this will alienate foreign capital, local élites and the emerging group of indigenous entrepreneurs; third, the government can nationalise the industry and look to Chinese and Soviet models but this will be expensive, cumbersome and wasteful of the country's human and material resources as the industry should remain 'low priority' for at least several decades; finally, the government can accept some degree of conflict and attempt to maximise benefits while minimising disbenefits. It is obviously hard to decide on philosophies or strategies and to reconcile things which are by their very nature contradictory. However, this essay advocates the last approach so that national and provincial governments have the opportunity to exercise some control over the industry and make it serve their purposes. Certainly there seem to be few advantages in allowing Papua New Guinea to become an appendage of the international tourist trade.

Acknowledgments

I am grateful for the comments, criticisms and advice of Steve Britton, Roger Lee, Gerard Ward, Peter Williamson and the staff of the Office of Tourism in Port Moresby who kindly welcomed, I think, an asp to their bosoms. At a more personal level I am indebted to Jim Griffin who made me ask questions, Wendy Alpers who persuaded me to become a tourist for a few hours, and Ian Douglas who gave me a chance to think.

References

Ala'ilima, L.P. 1977. Revising tourism in Western Samoa. Pp. 57-8 in B.H. Farrell (ed.), *The Social and Economic Impact of Tourism on Pacific Communities.* Center for South Pacific Studies, University of California, Santa Cruz.

Beier, U. 1977. Tourism in Papua New Guinea. (Paper read at seminar on 'Tourism and the Future: Coordination or Chaos?', October 1977, Townsville College of Advanced Education.)

Bryden, J.M. 1973. *Tourism and Development: A Case Study of the Commonwealth Caribbean.* Cambridge University Press, Cambridge.

Bugnicourt, J. 1977a. The new colonialism: tourism with no return! *Development Forum* 5(5): 1-2.

Bugnicourt, J. 1977b. Tourism: the other face. *Development Forum* 5(6): 8.

Bureau of Statistics Papua New Guinea 1978. National Account Statistics. *Statistical Bulletin,* 18 March 1978.

Burkart, A.J. and Medlik, S. 1974. *Tourism: Past, Present and Future.* Heinemann, London.

Central Planning Office 1975. *Fiji's Seventh Development Plan 1976-1980.* Government Printer, Suva.

Central Planning Office 1976. *The Post-Independence National Development Strategy.* (Papua New Guinea Government White Paper.) Central Planning Office, Waigani, Port Moresby.

Department of Labour, Commerce and Industry 1976. *Papua New Guinea: National Investment Strategy.* National Executive Council, Port Moresby.

Fraser, I.M. 1972. Participation in change — tourism, the two edged. *Study Encounter* 8(2): 1-8.

Gibbings, M.J. 1973. *Tourism in Papua New Guinea.* (Report prepared for the Papua New Guinea Department of Foreign Relations and Trade.) Department of Economics, University of Queensland.

Goulet, D. 1971. *The Cruel Choice: A New Concept in the Theory of Development.* Atheneum, New York.

Harcourt, T. 1977. The impact of tourist expenditure. *Yagl-Ambu,* 4: 174-90.

International Bank for Reconstruction and Development 1965. *The Economic Development of the Territory of Papua and New Guinea.* Johns Hopkins, Baltimore.

Lea, D.A.M. 1975. *Creating Convivial Spatial Structures. An Inaugural Lecture Delivered in Armidale, New South Wales on 13 October, 1975.* University of New England, Armidale.

Lea, D.A.M. 1977. The education of a tourist. (Paper read at seminar on 'Tourism and the Future: Coordination or Chaos?', October 1977, Townsville College of Advanced Education.)

Lipton, M. 1977. *Why Poor People Stay Poor: A Study of Urban Bias in World Development.* Temple Smith, London.

May, R.J. and Skeldon, R. 1977. Internal migration in Papua New Guinea: an introduction to its description and analysis. Pp. 1-26 in R.J. May (ed.), *Change and Movement: Readings on Internal Migration in Papua New Guinea.* Papua New Guinea Institute of Applied Social and Economic Research in association with the Australian National University Press, Canberra.

Naipaul, V.S. 1969. *The Middle Passage: the Caribbean Revisited.* Penguin, Harmondsworth.

Nayacakalou, R. 1972. Investment for tourism in Fijian land. *Pacific Perspective* 1: 34-7.

Persistently Empty Heart

Chapter 14

Resource Management in Arid Australia

Implications for the Development of the Conservation Movement in Australia

*R. L. Heathcote**

A definitive history of land settlement and associated resource management in the arid zone of Australia (Fig. 14.1) has not yet appeared but sufficient research has been done to show that settlement has been difficult, has still to be completed, and has reflected differences in environmental perception between settlers (people, institutions or companies) and officials (whether of the executive, judiciary or administrative branches). In Roberts' (1924) survey of Australian settlement to 1920 and the regional studies of King 1957; Meinig 1962; Heathcote 1965a, 1977; Blainey 1966; Kelly 1966; Duncan 1967; Powell 1970; Jeans 1972; and Williams 1974 those differences have been shown to influence resource management significantly. For the pastoral settlers in particular a 'pastoral ethic' has been recognised — attitudes and actions which have enabled a difficult environment to be exploited with at least partial and short-term success (Heathcote 1969a).

The problems which have faced resource management in the arid zone and the apparent conflict between settlers and officials (and the existence of a 'pastoral ethic' which seems to have been part of that conflict) have a wider relevance, not least to the current debates on the conservation of national resources. It will be argued that the

* *R.L. Heathcote is Reader in Geography, Flinders University of South Australia. After graduating from The University of London and The University of Nebraska, he took his doctoral degree in the Department of Geography, Research School of Pacific Studies, The Australian National University, in 1963.*

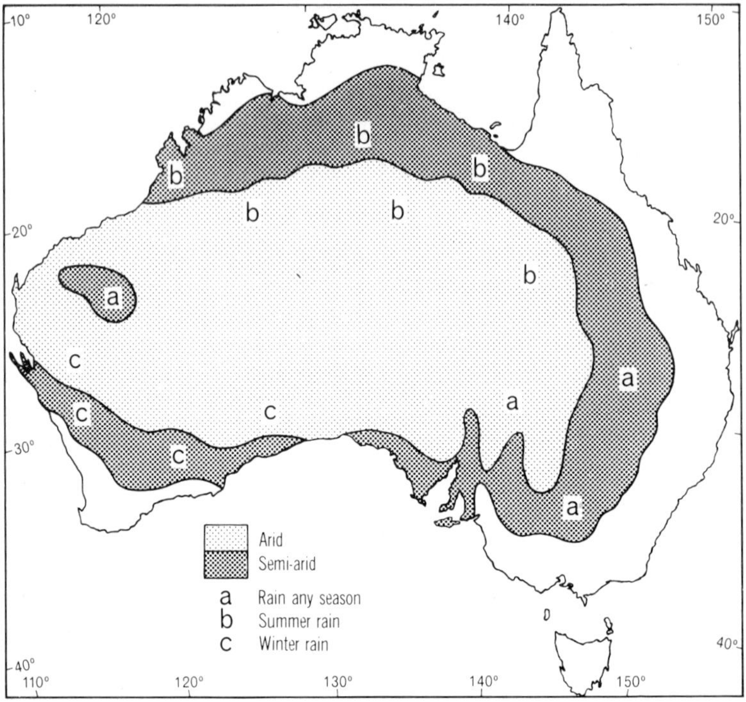

Figure 14.1 The arid zone of Australia (after Meigs 1953).

process of arid land settlement in Australia polarised attitudes to resource management between settlers and governments; that this resulted in a shift in the governmental role in resource management from that of real estate agent to that of steward for the public estate; and that this polarisation of interests and the redefinition of the governmental role were integral to the historic process leading to the current concern for environmental quality and the conservation of national resources. Before considering these hypotheses, however, the character of the current concern for the conservation of the environment, the historical and geographical contexts of arid land settlement and the nature of the pastoral ethic need to be outlined.

Concern for the Conservation of the Environment

As Glacken (1967) has shown, the origins of concern in Western societies for the impact of human activity on the environment may be traced at least to classical antiquity, but the scope of both the philosophies and the membership of what has come to be recognised as the conservation movement has broadened considerably since Rachel Carson's (1962) documentation of the effects of indiscriminate use of insecticides on bird life. The 1960s and early 1970s have seen a vast amount of literature published, extensive lobbying by conservation groups, legal conflicts and legislation requiring 'environmental impact' assessments of major official resource development projects (Ehrlich and Ehrlich 1970; Nicholson 1970). In Australia the policies and publications of the Australian Conservation Foundation (1971) and independent publications (such as Costin and Frith 1971; Marshall 1966; and Strom 1972) have generally mirrored the overseas concern.

At the risk of over-simplification, the main philosophy of the conservation movement, as illustrated from the sources mentioned, appears to be centred around the following beliefs:

(a) The earth is a finite resource for mankind.
(b) Previous human activity, particularly in Western technology-orientated societies, has reflected the belief in unlimited earth resources and has led to the wastage of irreplaceable assets.
(c) If mankind is to survive on earth, future human activity will have to balance its demands against apparently dwindling irreplaceable resources, mainly by attempting both to reduce its demands and to supply its needs on ecological principles of sustained yield rather than short-term exploitation.
(d) To achieve that balance individual freedom of resource management may have to be restricted by governmental controls if individual management conflicts with management of national resources for the benefit of the general public.

Most of the contemporary debate seems to focus upon the principles involved and means to achieve this final belief.

Arid Land Settlement

The arid lands of Australia were occupied from the mid-nineteenth century by European settlers moving out from the established centres in the more humid southern Australia. The arid lands presented to those settlers a harsher environment in terms of scarcity of water, paucity of vegetation and variability of the seasons than any areas previously experienced on the continent. This harshness, together with remoteness from the continental hearths of European settlement, reinforced the difficulty (already encountered in the humid coastal areas) of the contrasting life-style of the indigenous population which often led to conflict (Rowley 1972). In effect, the new settlers moved into an alien environment, knowledge of which initially was scanty and often erroneous, and that knowledge subsequently was only slowly and often painfully improved.

Despite these problems, however, the new settlement of the arid lands was encouraged by the contemporary mental climate in Western society, in which man was considered able to dominate and improve on nature — whose environmental challenges only awaited the application of human ingenuity for their resolution. Such philosophies of the inevitability of progress were paralleled by political ideologies which viewed the settlement of a nation's lands as a necessary goal for national development. This development was expected to result from the application of labour and capital to the land in a context of *laissez-faire* economics.

The application of human ingenuity did in fact make possible occupation of some of the new lands by the invaders in the latter half of the nineteenth century. Superior weaponry enabled the subjugation of the indigenous population. Innovations in land transport and communications (principally railways and telegraph/telephone systems) as well as the choice of durable products (wheat and wool) and refinements in the preservation of perishables (especially meat chilling) broke the 'tyranny of distance' by making possible the links between remote producers and global markets. Mechanisation of labour reduced the costs of land clearance, the sowing and harvesting of crops, and the shearing of sheep, while innovations in well-drilling technology and water-lifting devices (particularly the windmill pump) tapped the underground waters. Finally, new strains of crops and livestock produced more drought-resistant and higher-yielding varieties (of wheat and barley for the crops, the Merino crosses for the sheep and, more recently, of Zebu crosses with British breeds for the cattle).

The settlers, however, faced more than mere environmental challenges: the difficulties facing arid land settlement stemmed in part from the official land settlement policies themselves.

Official Land Policies

During the 1860s the colonial governments in New South Wales, Queensland, South Australia and Victoria introduced legislation for the settlement of their remaining unalienated lands, at least 60 per cent of which were in the arid zone. Except for small public reserves (mainly for communications), such lands were to be sold at fixed prices (part down-payment and part on credit over several years) to individuals, in units thought large enough to support a settler and his family. The various Acts were intended to place on the land settlers described in the legislative debates as an industrious and honest yeomanry. Extensive pastoral use, which was occupying much of those lands, either illegally or on pastoral leases, was regarded as but a temporary encumbrance which must in time give way to the proposed family farms.

From hindsight, three assumptions are apparent in the legislation — assumptions which were to pose major obstacles to the effective occupation of the arid zone and to create conflicts between settlers and governments. The first assumption was that all land was of uniform quality and productive potential, hence the standard size of settlement units within the colonies (varying admittedly somewhat *between* the colonies and with the accuracy of the land survey even *within* the colonies but generally within the range of 320-640 acres (130-259 ha) and the standard price of £1 per acre ($A4.94 per ha), only a portion of which was required as down-payment. The second assumption was that the settlement unit was sufficient to support an average settler and his family from agriculture. The third assumption was that disposal of the land was to be a permanent transaction which would result in bona fide settlement, i.e. occupation and use of the land as originally intended in the legislation.

For the pastoralists, the subsequent century of struggle to defend their resource management systems in the arid zone against this officially-backed agricultural encroachment has left them apparently victorious, as they still occupy the largest areas of Australia's arid zone (Heathcote 1969b). In that victory the development of a pastoral ethic seems to have played its part.

The Pastoral Ethic

The pastoral ethic, as originally defined (Heathcote 1969a), has three characteristics. First, the pastoral settlers saw themselves as independent-minded, self-made men; they were suspicious of their fellows and unwilling to accept either advice from, or control of their activities by, governments which were generally regarded as remote and unsympathetic. Second, the commercial motive dominated resource management. Profits, whether from livestock or from land sales, were all-important; short-term gains (opportune use) seemed to pay higher dividends than management aimed at conserving the 'evanescent' resources of fodder and water. Third, the settlers exploited their public image as pioneers, especially in extracting help from governments at times of environmental stress — particularly droughts.

In a broader view, however, it seems likely that this ethic was not confined to the pastoralists but might also be found amongst other types of potential settlers in the arid zone. Nevertheless, the general process of arid zone settlement reflects the characteristics of the pastoral ethic in the way in which attitudes to resource management polarised between settlers and governments.

Bona Fide Settlement? The Polarisation of Attitudes

Official policies intended the land to be used by the original purchaser but for many of the settlers land was a speculative commodity rather than a potential home. Reading the debates on the numerous amendments to the original Acts introduced by governments anxious to close loopholes against land speculation leaves the impression that officialdom was either incredibly naive about, or hoodwinked by, the land speculators' activities. Whatever the reasons, the loopholes persisted and were exploited. As a result, alongside the frequent need to amend the legislation, governments were forced to revoke titles for unfulfilled covenants and to police the activities of the settlers to an extent unforeseen in the original legislation (King 1957; Heathcote 1965a; Duncan 1967; Roberts 1968; Powell 1970). Such police action itself polarised officials and the speculative settlers.

Moreover, governments forgot that they were legislating for land settlement within a context of *laissez-faire* resource management in a commercially-orientated economy. The land *was* a commodity, which had to have a commercial attraction before many potential buyers would be even interested, and there was no guarantee that their interests would be other than short-term. In the half-century after the new legislation, the largest profits were coming from the traditionally large pastoral leaseholds rather than the new small farms, and while there is no doubt that the farms were attractive to small capitalists it is not certain whether their attraction lay in their agricultural or speculative potentials. Certainly in the arid zone, the evidence of the last 125 years has shown that for the pastoral properties and to a lesser extent for the farms, the larger the enterprise (whether in land or capital) the better the chance of profit and even survival in the long term (Barnard 1962: 383-400). Further, because of deteriorating productivity (which will be examined later in the essay) there has been successful pressure from settlers to expand the maximum permitted size of holdings to try to maintain their income levels (Anderson 1970; Heathcote 1965a; Lewis 1963).

To acquire, develop and operate the necessary areas required more capital than the average new settler could provide. *Extra* land was always more costly to obtain than the original block because of the need to buy out speculative owners, or to bribe dummy owners, or to pay the bona fide neighbours an attractive bonus to quit.

Substantial capital was needed by the dry farmer to provide the mechanical equipment required to cultivate the much greater area of his wheat farm in low rainfall country, and by the pastoralist to purchase a herd large enough to ensure himself against drought losses, to fence his range and put down the wells. While there is some evidence that speculative capital was available for dry-farming settlement, by far the largest amounts of overseas capital (up to the 1960s mainly British but thereafter including American) went into the pastoral properties of the arid zone (Bailey 1966; Barnard 1962; 322-39, 383-400; Butlin 1962, 1964). As a result company, non-family, land-ownership began early and has contined to be a significant feature of arid zone settlement (Barnard 1962: 411-71; Heathcote 1965a; Kelly 1966; Wells and Bates 1969). In the dominant (pastoral) settlement such ownership has been reported as reaching 45 per cent of the total (Rose 1954). The need for capital has weeded out the settlers of limited means, for whom, at least in part, the land was intended, but official opposition to any tendency towards monopoly of resources has remained.

As experience of the administration of the arid lands grew, officialdom became increasingly concerned at the attitude of some of the settlers. As early as 26 May 1869 the Queensland Secretary for

Public Lands claimed in Parliament that: 'Formerly a comparatively trifling sum was sufficient to keep a squatter's family; but now-a-days they must have carriages to ride about in, and town-houses to live in, and all the like of that extravagance'. The parliamentary debates and the long discussions about the optimal size of holdings showed that, like the rest of the community, the expectations of the settlers were increasing over the years, particularly the expectations of profits with which to improve the settlers' material welfare. Looking back from 1928 the Queensland Land Administration Board (1928: 11) in its first report noted:

> In the early days of Australian settlement a family might settle on the land, produce almost all its own requirements, and earn in actual money a very small income, which would be expended in buying articles that the farm could not produce. To live, rather than to earn or produce for the use of others, was the dominating purpose. This simple mode of life no longer satisfies, and primitive settlement of this kind now definitely belongs to the past.

The problem arose partly from the alternative amenities associated with urban employment and complaints about the social penalties of life in the arid zone were appearing at the turn of the century (Heathcote 1965a: 155). The rural-urban population drift (off both pastoral and farm properties) which seems to have been associated with this problem is well known; perhaps less well known is the extent to which this shift was a deliberate resource management strategy. For the boom years of the 1960s one experienced researcher noted that 'the attitude of some who commence woolgrowing in such districts is to make sufficient money to move into more favoured areas, where droughts occur less frequently' (Moule 1969: 181). For whatever reason, however, the turnover in land ownership and the slow attrition of settlers have continued and while it has been easy for officials, sitting in their city offices, to decry the process, it has been much less easy for them to do anything about it.

Sufficient Land for the Settlers?

Land policies were initially based on the assumption that the government's main function was to act as a real estate agent for the transfer of public lands to private ownership. However, the acreage made available to each settler was soon recognised to be inadequate and each government found that its involvement in land transfers not only lasted much longer than originally anticipated, but also became far more complicated. Instead of merely transferring land titles once, officials frequently found themselves dealing with the same block of land several times, either because a succession of settlers failed to comply with the stipulations for ownership, causing the title to revert to the government, or because illegal activities required the title to be revoked and the block to be reopened for settlement. The costs of these processes added significantly to the administrative burden and inflated the bureaucratic machinery (Roberts 1924). Perhaps the most spectacular example was the Marginal Lands Scheme in southeastern Australia by which farms in the semi-arid zone, hit by economic depression and the effects of soil erosion and crop failure from droughts in the 1930s and early 1940s, were repurchased by State governments with $A8 million of Commonwealth funds and enlarged by amalgamation for development as mixed crop and livestock farms (Proctor 1940; Heathcote 1969c). Similar 'rural reconstruction' is still going on (Williams 1976).

The need to enlarge the original block sizes not only absorbed bureaucratic man-hours and public funds, it also forced some rethinking of the official role in land settlement. The agitation from disgruntled settlers and their elected representatives, and the lobbying by interest groups, often resulted in an official inquiry — usually by a Parliamentary Select Committee or Royal Commission — whose report was the subject of further parliamentary debate before new legislation resulted. These inquiries and debates involved substantial inputs of fact and opinion (running to hundreds of pages of evidence) from officials, settlers and other interested parties. Here was a forum wherein the intent and achievement of official settlement policies were investigated (Heathcote 1963: Appendix I). It is in these inquiries that perhaps the first evidence of a change in the governments' role appeared for, as soon as officialdom accepted that there was a case to reconsider the sizes of the original blocks, they took on themselves responsibility for the success or failure of land settlement insofar as it could be attributed to block size and, by implication, the economic return therefrom.

Most of the inquiries in fact faced *inter alia* the problem of the optimal block size for settlement. The usual solution was to set up a classification of land by its productivity, actual or potential, and either allocate sizes to provide theoretically equal returns from land under a common use (e.g. possible grazing densities for pastoral lands) or to differentiate the possible uses (e.g. distinguishing between land to be grazed from land for dry-farming or for irrigation farming). Whichever method was used to determine block size, greater government involvement in, and responsibility for, resource appraisal and management was implicit. Once land classifications were written into land regulations, governments had begun to erode the settlers' freedom of action in resource management.

From the earliest period pastoral occupation had been on terms which allowed *only* pastoral use because the 'depasturing' or grazing licences and later the pastoral leases were seen as only interim measures — means by which a portion of the whole resource could be used for a limited period of time. However, the limitation of use and the specific exclusion of agriculture have continued up to the present — reinforced by official attempts to classify the arid resources and allocate them to specific users (Heathcote 1969b). The creation of the various 'agricultural areas' and 'irrigation areas' was a parallel effort to rationalise resource use through land settlement regulation (Jeans 1972; Powell 1970; Williams 1974).

Concern for resource management policies had been in one sense forced upon governments because of their inability to dispose of the arid zone to private ownership. In the 1970s the Crown is still owner of the vacant lands (some 27 per cent) of the arid zone and retains ultimate ownership of the leased (mainly pastoral) lands (some 65 per cent) and a portion of the reserves (some 7 per cent but excluding those areas in process of allocation to the Aborigines). As ultimate owners, therefore, governments still have the major responsibility for resource management policies in the arid zone; and the experience of dealing with those problems and the growing activities of the conservation lobby seem to be reinforcing pressures to retain that responsibility, rather than to continue the attempt to delegate it to others.

Arid Land Quality and the Conservation Movement

Most official land policies after 1860 were basically anti-monopolistic. The intention was to spread resource use among as many users as possible. The arid zone, however, was particularly vulnerable to monopolistic control of resource use because of the uneven occurrence of the scarcest but basic resource — water — and the general practice of individual rather than communal ownership of resources. As the first pastoralists were quick to demonstrate, where fodder was widely if thinly spread but water limited, possession of the few water supplies gave effective control of the surrounding dry rangeland. The discovery of underground waters did not significantly alter the situation because the depth of the aquifers (over 1,000 m on average) required substantial capital to develop the resource (as noted previously) and the few capitalists or companies had an immediate advantage over the average settler. Governmental reaction to both situations was in the first case to reserve, wherever possible, a proportion of the water supplies for public use and itself put down wells and reserve them as public watering-places (Heathcote 1965b).

This action illustrated the need to adapt traditional British water law and management to arid zone conditions, a process still under way and which has recently been the subject of investigations by the Australian Water Resources Council (Clark and Renard 1972). Implicit has been the need for arbitration between competing users, a process increasingly involving government — not merely as arbitrators but to a mounting degree as plaintiffs. This latter situation arose particularly in relation to irrigation developments along the Murray River and disputes between New South Wales, South Australia and Victoria about ownership of the river (Rutherford 1964). The solution here was the creation of an apparently independent River Murray Commission but its independence has been questioned recently and the potential for dispute apparently remains (Frith and Sawer 1974: 272). This potential, if it exists, is an argument for national control of this resource, for an effective arbitrator *between* the States and resource management on behalf of national rather than regional interests.

Implicit in nineteenth century philosophy on land settlement was the belief that the settlement process physically improved the land and increased its value as a commodity. Physical improvement resulted from the investment of labour and capital in transforming the environment, releasing the inherent fertility of the soil and ameliorating the climate by tillage and the manipulation of the vegetation. Increased land value accrued partly from locational advantages associated with intensification of land use on the edge of previously settled areas and partly from the 'rent' resulting from the inputs of labour and capital to bring the land into production (Heathcote 1963). Such positive changes in the arid lands, however, have been periodically negated by droughts, which caused heavy crop and livestock losses (Foley 1957; Lovett 1973), and evidence of long-term environmental deterioration of both grazing and farming areas.

Faced by drought-induced production losses, settlers have applied for official relief. Governments traditionally have been sympathetic and there has been a long history of relief measures, particularly in the semi-arid areas where humid land resource management was introduced with minimal modifications (Heathcote 1969c). Such relief has usually been provided with the aim of restoring the settler to his position before the drought — implying that there exists an ecological balance in the arid lands which has been only temporarily upset. Evidence is increasing, however, that the upset may be long lasting.

Evidence of environmental deterioration came from the reduction of livestock carrying capacities on the arid ranges and from soil erosion on the dry-farming areas. By the turn of the century in eastern Australia the consensus was that the capacities of the ranges had been reduced by half from their peaks in the 1890s and fears were being expressed about the long-term effects of over-grazing on the capacity of the ranges to support intensive pastoral settlement (New South Wales 1901; Queensland 1897). Similar concern appeared in Western Australia in the 1930s (Western Australia 1963) and the Northern Territory in the 1960s (Condon *et al.* 1969), and official fears for a general 'desertification' have been reinforced recently (Australia 1977; Laut *et al.* 1977; South Australia 1976; Williams *et al.* 1977). Such concern, particularly that private mismanagement, resulting from over-optimistic assessment of long-term environmental productivity, had caused significant deterioration of a resource basically still owned by the general public, was paralleled by the arguments for the official soil conservation measures introduced as the result of massive erosion from the semi-arid farm lands in the 1930s and early 1940s in southeastern Australia (Campbell 1948; Williams 1974). For both the vegetation of the ranges and the soils of the farms, governments adopted a role as guardian of a public and national resource under threat from apparent private misuse, whether accidental or deliberate. The pressure to adopt this role, which is now recognised as a fundamental principle of the conservation movement, developed early in the twentieth century and has increased significantly with the continued experience of the environmental impact of land settlement in the arid zone.

Conclusion

Land settlement in the arid zone of Australia since the mid-nineteenth century has involved the introduction of alien settlers and settlement systems into an essentially hostile environment. Approximately two-thirds of the arid zone has been settled but this has been achieved only by the adaptation of the introduced systems of settlement and associated resource management, by administering the process of settlement long beyond the anticipated period of official involvement at a significant but as yet inadequately measured cost, and by a significant and possibly long-term modification of the arid environment.

Adaptation of the systems of land settlement and associated resource management has been at two distinct levels — that of the settlers in the arid zone and the governments in the humid zone. For the settlers, adaptation was vital. For many, their efforts were inadequate and they retired ruined from the fray; for some, who had no intention of staying, their 'retreat' was a strategic withdrawal after a successful speculation; for others, perhaps the minority, their adaptations proved financially rewarding. For governments, adaptation meant more office work and greater involvement in the whole process of land settlement. This resulted from the need to police the land regulations; to encourage the small landholders to survive the natural tendency towards amalgamation and larger units; to protect resources from private monopoly; and to protect those resources not yet finally allocated from misuse. This increasing official involvement seems to have produced concepts of resource management very similar to those implicit in contemporary conservation policies.

To suggest that the Australian conservation movement was born and nurtured solely in the arid zone would deny its international links and the broader national bases, which are now being documented (Powell 1976). But the adaptations of traditional policies of land settlement and resource management required to establish resource use in the arid zone in the last 125 years have illustrated conflicts and raised issues which are central to the present debate on the conservation of Australian resources. The extent to which this illustrates parallel and perhaps mutually reinforcing philosophical developments, or represents the origins of the current philosophies, is uncertain but would be worth investigation.

References

Anderson, J.R. 1970. A case study of spatial diversification of pastoral sheep holdings. *Review of Marketing and Agricultural Economics* 38: 137-41.

Australian Conservation Foundation 1971. *Policies of the Australian Conservation Foundation (as at June 30 1971).* Melbourne.

Australia, Department of Environment, Housing and Community Development 1977. *Experiences of Anti-Desertification Campaigns in Australia.* Canberra.

Bailey, J.D. 1966. *A Hundred Years of Pastoral Banking: A History of the Australian Mercantile Land and Finance Company 1863-1963.* Clarendon Press, Oxford.

Barnard, A. (ed.), 1962. *The Simple Fleece: Studies in the Australian Wool Industry.* Melbourne University Press, Melbourne.

Blainey, G. 1966. *The Tyranny of Distance: How Distance Shaped Australia's History.* Sun Books, Melbourne.

Butlin, N.G. 1962. *Australian Domestic Product, Investment and Foreign Borrowing 1861-1938/39.* Cambridge University Press, Cambridge.

Butlin, N.G. 1964. *Investment in Australian Economic Development 1861-1900.* Cambridge University Press, Cambridge.

Campbell, K.O. 1948. The development of soil conservation programmes in Australia. *Land Economics* 24: 63-78.

Carson, R. 1962. *Silent Spring.* Houghton Mifflin, Boston.

Clark, S.D. and Renard, I.A. 1972. *The Law of Allocation of Water for Private Use* (4 vols). Australian Water Resources Council Research Project 69/16, Melbourne.

Condon, R.W., Newman, J.C. and Cunningham, G.M. 1969. Soil erosion and pasture degeneration in central Australia. *Journal of the Soil Conservation Service of New South Wales* 25: 47-92, 161-82, 225-50, 295-321.

Costin, A.B. and Frith, H.J. (eds.), 1971. *Conservation.* Penguin, Melbourne.

Duncan, R. 1967. *The Northern Territory Pastoral Industry 1863-1910.* Melbourne University Press, Melbourne.

Ehrlich, P.R. and Ehrlich, A.H. 1970. *Population Resources Environment: Issues in Human Ecology.* Freeman, San Francisco.

Foley, J.C. 1957. *Droughts in Australia: Review of Records From Earliest Years of Settlement to 1955.* Commonwealth of Australia Bureau of Meteorology Bulletin No. 43, Melbourne.

Frith, H.J. and Sawer, G. (eds.), 1974. *The Murray Waters: Man, Nature and a River System.* Angus and Robertson, Sydney.

Glacken, C.J. 1967. *Traces on the Rhodian Shore: Nature and Culture in Western Thought from Ancient Times to the End of the Eighteenth Century.* University of California Press, Berkeley.

Heathcote, R.L. 1963. Historical changes in the appraisal of pastoral land resources. *Review of Marketing and Agricultural Economics* 31: 209-29.

Heathcote, R.L. 1965a. *Back of Bourke: A Study of Land Appraisal and Settlement in Semi-Arid Australia.* Melbourne University Press, Melbourne.

Heathcote, R.L. 1965b. Changes in pastoral land tenure and ownership: an example from the western division of New South Wales. *Australian Geographical Studies* 3: 1-16.

Heathcote, R.L. 1969a. The pastoral ethic: a comparative study of pastoral resource appraisals in Australia and America. Pp. 311-24 in W.G. McGinnies and B.J. Goldman (eds.), *Arid Lands in Perspective Including AAAS Papers on Water Importation Into Arid Lands.* The American Association for the Advancement of Science, Washington and University of Arizona Press, Tucson.

Heathcote, R.L. 1969b. Land tenure systems: past and present. Pp. 185-208 in R.O. Slatyer and R.A. Perry (eds.), *Arid Lands of Australia.* Australian National University Press, Canberra.

Heathcote, R.L. 1969c. Drought in Australia: a problem of perception. *Geographical Review* 59: 175-94.

Heathcote, R.L. 1977. Pastoral Australia. Pp. 252-88 in D.N. Jeans (ed.), *Australia: A Geography.* Sydney University Press, Sydney.

Jeans, D.N. 1972. *An Historical Geography of New South Wales to 1901.* Reed Education, Sydney.

Kelly, J.H. 1966. *Struggle for the North.* Australasian Book Society, Sydney.

King, C.J. 1957. *An Outline of Closer Settlement in New South Wales: Part 1 The Sequence of the Land Laws.* New South Wales Department of Agriculture, Sydney.

Laut, P. *et al.* (comps) 1977. *Environments of South Australia: Province 8 Northern Arid.* Division of Land Use Research, Commonwealth Scientific and Industrial Research Organization, Canberra.

Lewis, J.N. 1963. Is the concept of the home maintenance area outmoded? *Australian Journal of Agricultural Economics* 7: 97-106.

Lovett, J.V. (ed.), 1973. *The Environmental, Economic and Social Significance of Drought.* Angus and Robertson, Sydney.

Marshall, A.J. (ed.), 1966. *The Great Extermination: A Guide to Anglo-Australian Cupidity, Wickedness and Waste.* Heinemann, Melbourne.

Meigs, P. 1953. World distribution of arid and semi-arid homoclimates. Pp. 203-9 in *Reviews of Research on Arid Zone Hydrology.* United Nations Educational, Scientific and Cultural Organization, Paris.

Meinig, D.W. 1962. *On the Margins of the Good Earth: The South Australian Wheat Frontier 1867-1884.* Rand McNally, Chicago.

Moule, G.R. 1969. Research and the woolgrower. *Australian Journal of Science* 32: 176-84.

New South Wales 1901. Royal Commission to inquire into the condition of the Crown tenants (appointed 11 August 1900): Report, summary and minutes of evidence, appendices and returns. *Votes and Proceedings of the Legislative Assembly,* 1901(4): 131-1183.

Nicholson, M. 1970. *The Environmental Revolution: A Guide for the New Masters of the World.* Hodder and Stoughton, London.

Powell, J.M. 1970. *The Public Lands of Australia Felix: Settlement and Land Appraisal in Victoria 1834-91, with Special Reference to the Western Plains.* Oxford University Press, Melbourne.

Powell, J.M. 1976. *Environmental Management in Australia, 1788-1914. Guardians, Improvers and Profit: an Introductory Survey.* Oxford University Press, Melbourne.

Proctor, M.L.R. 1940. Marginal land: South Australia and New South Wales compared. *Australian Geographer* 3(8): 16-31.

Queensland 1897. Report of the Royal Commission on land settlement; together with the minutes of proceedings, minutes of evidence taken before the Commission, and appendices. *Votes and Proceedings of the Legislative Assembly*, 1897(3): 869-1242.

Queensland 1928. First annual report of the Land Administration Board. *Parliamentary Papers*, 1928(2): 1-59.

Roberts, S.H. 1924. *History of Australian Land Settlement 1788-1920*. Macmillan and Melbourne University Press, Melbourne.

Rose, F.G.G. 1954. The pastoral industry in the Northern Territory during the period of Commonwealth administration, 1911-53. *Historical Studies Australia and New Zealand* 6: 150-72.

Rowley, C.D. 1972. *The Destruction of Aboriginal Society*. Penguin, Melbourne.

Rutherford, J. 1964. Interplay of American and Australian ideas for development of water projects in northern Victoria. *Annals of the Association of American Geographers* 54: 88-106.

South Australia 1976. *Vegetation Clearance in South Australia, Report of the Interdepartmental Committee*. Government Printer, Adelaide.

Strom, A.A. 1972. Conservation as a goal. Pp. 52-66 in J.A. Sinden (ed.), *The Natural Resources of Australia: Prospects and Problems of Development*. Angus and Robertson, Sydney.

Wells, J.M. and Bates, W.R. 1969. Changes in farm business organisation in Australia. *Quarterly Review of Agricultural Economics* 22: 53-65.

Western Australia 1963. *Report of the Pastoral Leases Committee*. Government Printer, Perth.

Williams, M. 1974. *The Making of the South Australian Landscape: A Study in the Historical Geography of Australia*. Academic Press, London.

Williams, M. 1976. Planned and unplanned changes in the marginal lands of South Australia. *Australian Geographer* 13: 271-81.

Williams, O.B. *et al*. 1977. *Gascoyne Basin (Case Study on Desertification)*. [Presented by Government of Australia to the United Nations Conference on Desertification.]

Chapter 15

From Vision to Pipe Dream

Yet Another Northern Miss

*G. J. R. Linge**

> We may set aside the vision — or the pipe-dream — of a great rural population in the north, but it still remains incumbent, morally and materially, that Australia should do what can be done to make it more productive. At best that may not be very much on a world scale, and it will not be done by blindly rushing ahead with grandiose schemes; but given the intractability of this vast and harsh land, it may yet be a great achievement. (Spate 1968: 165.)

Australians seem to need catch-cries which, by dint of repetition rather than by depth of reasoning, can become ingrained in the very substance of the body politic. In the fullness of time origins become obscure, definitions distorted and contexts confused so that no one can be quite certain what the terms really mean although they remain in common coinage. In the brave new world after World War II, for example, 'decentralisation' tripped lightly off many a tongue advocating industrial development of country towns (to cite but one meaning); they may not have been stilled but perhaps expectations would have been lessened had their owners been aware of the origins and activities of the widely supported but largely ineffectual

* *G.J.R. Linge is Professorial Fellow in the Department of Human Geography, Research School of Pacific Studies, The Australian National University. After graduating from the London School of Economics and taking his doctorate at The University of New Zealand, he joined the Department of Geography under Professor Spate as a Research Fellow in 1959.*

Decentralisation Leagues formed in Victoria and New South Wales during the 1880s. The 'White Australia' and 'populate or perish' syndromes, which likewise had their origins in the nineteenth century, have also passed through several phases. During the 1920s, in particular, geographers were dragged into these controversies, with the estimates of Australia's maximum population 'carrying capacity' made by Griffith Taylor (65 million) and Ellsworth Huntington (15 to 20 million) being much used — and abused — in the popular polemic. About the same time another attempt was made by the German geographer, Albrecht Penck, who calculated that Australia could produce food enough to support 480 million people. Although in many respects this was much the most 'scientifically-based' estimate, it assumed among other things that everyone would be living at a subsistence level. Perhaps this is an instance where if the expert had actually looked around this country his sums may have taken some account of the Australian materialistic expectation. Daneš, a visiting geographer from Central Europe at this time, was certainly aware of this factor (see Jennings' essay in this volume).

In various ways these catch-cries also form themes in yet another oft-heard Australian call, 'develop the north'. The arguments — part guilt, part fear — have been well summarised by Spate himself (1968: 153-65). He noted the 'brute economics' exposed by Bruce Davidson in *The Northern Myth,* the failure of some of the post-war schemes (like rice-growing at Humpty Doo), and the uncertainty surrounding others (such as the irrigation scheme on the Ord River which, ten years later, must be deemed another débâcle). But no sooner had Spate sounded the warning, quoted at the head of this essay, about grandiose schemes than another was appearing on the horizon. This episode, the rise and fall of the concept to promote the Pilbara region of northwest Western Australia as a major industrial complex, on the one hand reveals much about the problems of developing and populating the northern part of this continent while, on the other, focuses attention on some of the basic dilemmas confronting Australia as a nation.

The Pilbara Saga

Although the existence of large iron ore deposits in the Pilbara was documented in the 1880s they were excluded from estimates of the nation's reserves because it was thought improbable that minerals in

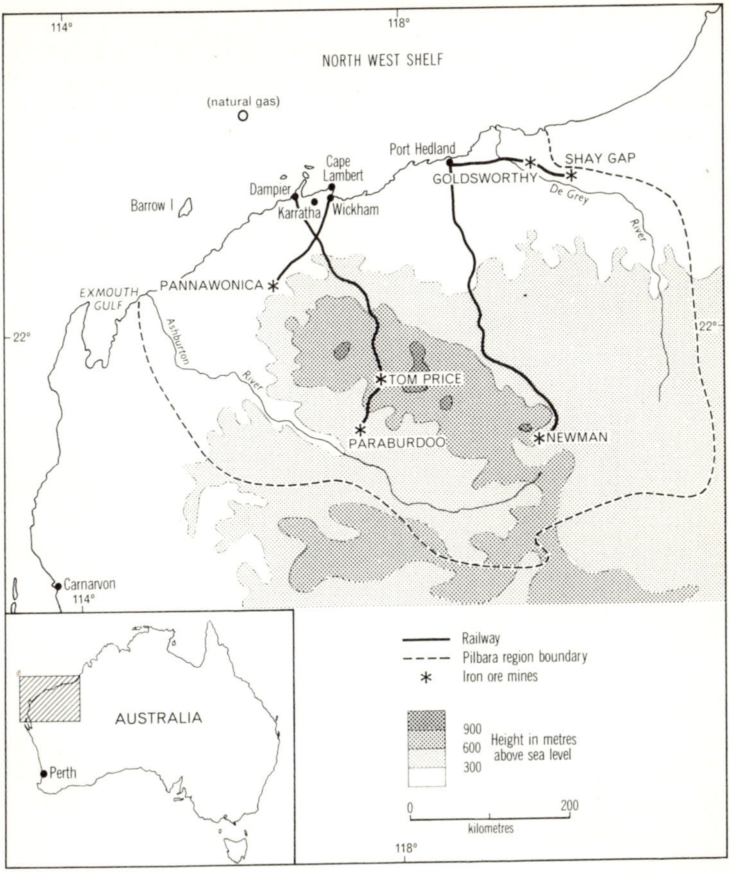

Figure 15.1 The Pilbara region, Western Australia.

such a remote, climatically harsh and sparsely populated part of the continent (Fig. 15.1) could ever be brought into production. They continued to be excluded when in 1938 the Federal Government imposed an embargo on iron ore exports to conserve supplies (then thought to total no more than 250 million tonnes) for Australia's own iron and steel industry.

Thus the mineral wealth of the Pilbara lay undisturbed until the early 1960s, when reserves were reassessed and the export embargo lifted. Apart from this legal change, however, other circumstances were also propitious. Several countries were looking for reliable, large volume supplies of iron ore; in Japan, for instance, plans were in hand to raise annual crude steel output from twenty to ninety million tonnes between 1960 and 1970. Moreover technology in several fields — ship design, port construction, railway building, materials handling and mining — had developed to the point where ore could be produced at a rate that was sufficient to offset the high capital and running costs of operating in such a remote area. Finally, international managerial and financial expertise was available, and could be brought together in consortia, to organise and support projects of the required scale.

It is probably fair to say that in the early stages the Western Australian government, which has full legal control of mineral exploration and development throughout the area, regarded the opening up of the Pilbara as a means to an end rather than as an end in itself. The granting in the mid-1960s of leases and concessions (which has led so far to four large-scale iron ore mining and exporting operations) provided the basis on which to open up a virtually uninhabited part of the State, created an additional source of revenue through royalties and payroll taxes, and suggested a basis for subsequent development of mineral processing and manufacturing activities. It also provided a psychological boost to a State which, like the others, was being allocated federal funds on the basis of its share of the Australian population (7.1 per cent in 1961) but which, in addition, was receiving supplementary grants in recognition of the costs involved in servicing a sparse rural population scattered over nearly one-third of the continent. A definite stigma attached to being a 'claimant' State: it was tantamount to being labelled a poor relation of the others.

As far as can be ascertained no development strategy for the Pilbara had been considered let alone set down on paper. Possibly one reason was that the initial grounds on which the first two mines were brought into operation in 1966 were export contracts for less than five million tonnes of iron ore per year. Moreover, the mine, townships, railway and shipping facilities associated with each of these developments were to be constructed independently in separate areas 200 kilometres apart and it was probably difficult for anyone to envisage the potential or the problems of developing the Pilbara as a region. It is also the case that the State simply did not have the financial resources to build the infrastructure for these ventures, especially within the time horizons on which the companies had predicated their contractual obligations. Nor was it possible for the State government to borrow funds to provide the infrastructure since all public borrowings by the State and Federal governments are co-ordinated by the Australian Loan Council which, under the conventions then operating, assessed the borrowing capacity of the States on the basis of their population. Yet at the same time the companies, anxious to secure access to the iron ore deposits, were prepared to finance not only the infrastructure directly required for their mining operations but also community facilities like houses, hospitals, schools, swimming pools, police stations and post offices.

By 1969, when the third iron ore mining and exporting project came into operation and others were at the planning stage, the State government was beginning to give some thought to broader development issues. For instance the possibility of a uniform electric power system for the region was mooted (but not proceeded with) and consultants were commissioned to prepare a Pilbara Regional Development Study. Soon after this report was submitted in May 1971, substantial reserves of natural gas were discovered on the North West Shelf, about 140 kilometres off the coast from Dampier, adding a new dimension to the potential of the region. Meanwhile at State elections in March 1971 the ruling Liberal and Country League-Country Party coalition had been replaced — after twelve years in office — by a Labor government. Whatever the reasons may have been, the consultants' report was never made public.

In 1972, about the time the fourth mining and exporting operation commenced production, the new State government produced a booklet, *The Pilbara — A Development Concept for the 1970s,* which was, in effect, a broad outline of a regional production complex based on natural gas, iron ore and salt. This envisaged the expenditure of $A5,800 million over a decade to provide an integrated complex of industrial activities, a new town of 130,000 people, a 4,400 megawatt power station, a water supply, and port facilities to handle twenty-five million tonnes annually. The concept outline argued that such a project would not get off the ground if the companies involved were expected, as they had been in the 1960s, to finance and construct almost the whole of the infrastructure; it thus proposed that $A2,400 million of the capital outlay involved should come from the public sector and be funded 'completely outside the customary Commonwealth/State financial relations' or, in other words, outside traditional Loan Council guidelines.

As another coincidence, the Federal elections in December 1972 brought a Labor government into power (the first for twenty-three years) and this was sympathetic both to planning *per se* and to proposals for regional development. Within a matter of weeks the Pilbara concept became the subject of a joint Federal-State study and this eventually reported in August 1974. On the assumption that the infrastructure would mainly be funded by the public sector, it was found that a wide range of industries would be viable (with petro-chemicals having the greatest potential) but that steel making would only be marginally profitable and the production of alumina and aluminium (using bauxite from outside the region) would be unprofitable. In addition, of course, the export of unprocessed and beneficiated iron ore from existing and possibly new mines would continue. But the Liberal-Country Party coalition (returned to government at the State elections in April 1974) gave the report of this study group a low-key reception, presumably because of uncertainties engendered by the announcement of the Federal Labor government in October 1973 that it would take control of the distribution and marketing of natural gas from the North West Shelf. Since the relevant legislation explicitly indicated an intention to market gas at a uniform price throughout Australia, it removed much of the incentive for processing and manufacturing industries to locate in the otherwise high cost Pilbara region. Furthermore, circumstances on a wider stage were beginning to affect the financial and market assumptions on which the study had been based. The decision by the Organisation of Petroleum Exporting Countries to increase the price of oil in October 1973, inflation in Australia and elsewhere, the downturn in world business activity, and the opening up of iron ore deposits in other parts of the world all affected the prospects of the iron ore mining and exporting projects already operating in the Pilbara to say nothing of the notion of an industrial complex.

The gas masque

It is necessary to consider another chain of links in this tangled dance. The passage by both the Federal and State legislatures of the *Petroleum (Submerged Lands) Act* 1967 established joint machinery for the administration of offshore petroleum tenements. One group with areas administered under this legislation was the consortium led by the Woodside-Burmah Oil group which held the North West Shelf lease where natural gas — the energy base for the proposed Pilbara complex — was discovered in 1971. This lease was due to expire in March 1975 and there were some doubts about whether the Minister for Minerals and Energy in the Federal Labor government would renew the permit. Eventually in June 1975 the Western Australian government indicated its dissatisfaction with the Federal government's attitude by unilaterally renewing this and other exploration permits sought by Woodside-Burmah. But other uncertainties remained. One was the possibility, mentioned earlier, that the Federal government would use its more wide-reaching offshore powers under legislation passed in 1973 to purchase the gas at the well-head and act as both distributor and wholesaler, thus foreclosing some of Woodside-Burmah's commercial options. (A legal challenge by the States to this legislation was not resolved until December 1975, hence adding a further complication.) Another contingency was whether the Federal government would ban exports in order to conserve energy supplies for Australia's internal use and thus deprive the company of baseload contracts. And there also remained a series of queries relating to taxation liability and the application of the regulations about the use of overseas capital. To make matters worse, the leading shareholder in the Woodside-Burmah consortium was a subsidiary of Britain's Burmah Oil Company which was running into difficulties in 1975 and 1976 largely through over-commitment to the construction of a tanker fleet. Even while it was rationalising much of the remainder of its corporate empire, Burmah Oil nonetheless held on to its shares in the North West Shelf project until August 1976 when it sold them to The Broken Hill Proprietary Company Ltd. By raising the local shareholding in the project to 48 per cent this action helped solve another potential difficulty since Federal government guidelines require that about half the equity in undertakings of this kind be held in Australia.

These uncertainties were not cleared up until the latter half of 1977 when the companies concerned announced that a package of agreements, assurances and incentives (not all of which have been made public) negotiated with the Federal and State governments provided 'a satisfactory environment in which to commit the substantial funds required for the planning and definition stage'. After four years of indecision and suspense, the Federal government agreed to allow nearly three-quarters of the initial daily production to be exported, thereby implicitly recognising the need for a substantial baseload to make the development of an offshore gas field in water 120 m deep economically worthwhile, and tacitly recognising the unlikelihood of any such baseload arising in Australia in the foreseeable future. In effect by this decision, the Federal and Western Australian governments together switched off the Pilbara complex: 74 per cent of the power from its dynamo was to go overseas and 21 per cent was to be piped to the southwest of the State. A puny 5 per cent would be retained for use in the Pilbara itself.

Whether or not the North West Shelf project itself will proceed remains unclear. The companies are still at the feasibility study stage concerned not only with technical matters but also with finance and markets. An immense capital outlay (estimated in 1977 to be from $A2,500 to $A3,000 million) will be involved in developing this offshore gas field, piping the gas to an onshore storage and liquefaction complex, and shipping liquefied natural gas overseas. The scale of expenditure is brought home more clearly if it is realised that, at current rates of inflation, every *hour* the project is delayed adds about $A20,000 to its cost. Such schemes have to be conceived in terms of an operational life of at least twenty years under 'take-or-pay' contracts which may, in turn, present financial risks that are acceptable only to very substantial, established users.

Under the 1977 'package' it was also proposed that 7.1 million m^3 (21 per cent of daily output) should be piped 1,300 kilometres from Dampier to the southwest where it was expected that much of it would be used by plants converting bauxite into alumina. This involved the State Energy Commission in a commitment to spend $A2,000 million over twenty years to build and operate the pipeline but posed questions as to whether it would be able to raise the necessary capital. In November 1978 the Loan Council gave its blessing to this pipeline project enabling the Western Australian government to try to drum up overseas funds. There have also been questions about potential markets although a State government announcement in October 1978 that one of the two alumina refinery projects being opposed on environmental grounds would be allowed to proceed appears to have increased the range of possibilities.

Implications

The circumstances whereby the vision of an industrial complex in the Pilbara has faded into a pipe dream illustrate some of the problems of further exploiting the mineral resources of northern Australia. The view taken here is that development will continue to be mainly on the basis of individual companies opening up deposits to produce minerals for shipment to processing plants overseas or in the more densely settled parts of Australia. Although some processing may take place (such as the beneficiation of iron ore and the liquefaction of natural gas), the emergence in northern Australia of complexes incorporating manufacturing activities based on these minerals is extremely unlikely and, in terms of Australia's economic future, undesirable.

It is unrealistic to suppose that manufacturing will develop spontaneously in association with these northern mining ventures. The only feasible possibility may be the establishment of large-scale, integrated operations by the mining companies themselves or with their agreement and co-operation. Much of the output of any such complex would have to be exported. This would make the whole system unstable as it would be subject to considerable externally-generated uncertainties stemming from variations in business activity in other countries, competition from other sources, and decisions by governments around the world about exchange rates, tariffs and quantitative restrictions. Since these eventualities would almost certainly affect the various products at different times, it is likely that the complex would always operate at less than its maximum efficiency.

The development and operation of a complex would depend on decisions made by agencies in the public sector on the one hand and enterprises in the private sector on the other. Both within and between these groups there are likely to be differences in the time horizons adopted and in the attitudes to risk: whereas firms in the private sector might seek to recoup their investment within ten years, government agencies would generally take a much longer view. It is possible of course that there could be some co-ordination of investment decisions by the public sector in a particular region, but this supposes a degree of co-operation between Federal and State governments and the multiplicity of agencies involved that has seldom been achieved in practice. Even the various instrumentalities of a State government would, at any particular moment, have differing priorities depending on the finances available, the lead time required to construct a facility or provide a service, and spatial imbalances in demand and supply. In reality, moreover, it is probable that ways will nearly always be found to give priority to existing developed areas since these are where the demand already exists, where there is electoral and political muscle to back up these demands, and where there are institutional arrangements to handle the administrative and construction work involved.

The Pilbara saga also illustrates the complexities inherent in putting together legal, financial and administrative packages to form consortia with the integrity and skills necessary to attract investment funds from world money markets. These complexities are exacerbated because members of these consortia may have dissimilar, and shifting, priorities in relation to their international obligations and arrangements. Moreover, changes in technology would occur at different rates and at different times in the various industries in a particular complex and this would affect the relationships within and between the consortia involved. If the consortia were established on the basis of rigid contractual agreements about demand and supply between firms, innovation would probably be stultified because changes could disrupt the whole framework.

The development of an industrial complex implies considerable investment in facilities by the public sector. As indicated already, the four iron ore mining and export projects in the Pilbara (between them producing about eighty-five million tonnes annually) themselves had to provide much of the infrastructure. If the whole of a town, port or railway serves one particular enterprise, this method of development creates few problems. But when facilities are to be used by several producers as well as the community at large it could be argued that in equity they should be provided by the public sector. Yet it is difficult to see how, under present Federal-State arrangements, funds of the order required could be found to develop and maintain the infrastructure of even one small complex.

Another set of problems relates to the labour force of a complex in a remote area with a relatively hostile climate. During the development stage most workers would be employed by a few large capital-intensive enterprises which, being geared to meet tightly scheduled commitments, could be readily threatened by employees seeking improvements in wages and conditions. In the Pilbara region, for example, attempts to attract a stable work-force to the iron ore projects, and subsequent efforts to meet demands for improvements, have led to a very high wage and wage-related cost structure. Mining enterprises, having to operate predominantly in isolated places with extreme climates in many parts of the world, build such costs into their anticipated investment and operating expenditure. But processing and manufacturing industries could not absorb such high labour costs and compete on domestic or overseas markets. There are thus two reasons to suppose that private enterprise would have been reluctant to invest in the processing and manufacturing industries in the sort of complex envisaged for the Pilbara. First, most firms would have traded off the additional costs of transporting the raw materials (such as iron ore, natural gas and salt) to less remote and more populous parts of Australia against the lower direct and indirect labour costs. Second, most firms would have preferred the relative anonymity of being merely one among many employers in a loosely structured organic city of a million inhabitants like Perth rather than being deeply and directly involved in the constant worker: management tensions that would be likely in a planned town of say 150,000 people in a remote part of the continent.

The 1970s have seen a sustained debate about the need to 'restructure' the Australian economy in general and the manufacturing sector in particular. The need to expand exports of industrial products has become a central theme of government policy and, in turn, carries implications about increasing efficiency, updating technology and becoming more competitive internationally. Manufacturers have been adjured to concentrate their efforts on activities in which Australia has a comparative advantage. Although the results are far from conclusive, some models of the Australian economy currently being devised and tested suggest that the further processing of some metallic minerals falls into this category. Those who espouse causes like 'decentralisation' and 'northern development' (or fulminate about multi-national companies treating Australia as yet another of their colonial quarries) may see here an opportunity for governments to 'require' companies to establish more-advanced processing plants adding further value to mineral exports and creating local employment. The chances are, however, that the great majority of these plants (whether set up voluntarily or under duress) will be located near existing centres of population on the east coast or near Perth. There are, and will continue to be, some exceptions but the notion of a spate of industrial complexes sprouting in the bush is simply not on.

The wheel of fortune often turns quickly in northern Australia. In the short view it may seem that it was unkind to the Pilbara. In a longer view Australians may yet be grateful that it spun fast enough to discourage people from 'blindly rushing ahead' with another 'grandiose scheme'.

Acknowledgment

Acknowledgment is made to the very considerable assistance given during the preparation of an early draft of this essay by Mr S. A. Hohnen, Principal Assistant Planning, Department of Industrial Development, Government of Western Australia. The views expressed are, of course, solely those of the author.

Reference

Spate, O.H.K. 1968. *Australia*. Benn, London.